VEDIC MATHEMATICS
NEW HORIZONS
ADVANCE LESSONS

Published by :
Lotus Press Publishers & Distributors

VEDIC MATHEMATICS
NEW HORIZONS
ADVANCE LESSONS

DR. S.K. KAPOOR
Ved Rattan

Lotus PRESS
4735/22, Prakash Deep Building
Ansari Road, Darya Ganj,
New Delhi - 110002

Lotus Press : Publishers & Distributors
Unit No. 220, 2nd Floor, 4735/22, Prakash Deep Building,
Ansari Road, Darya Ganj, New Delhi- 110002
Ph.: 41325510, 98118-38000
• E-mail : lotuspress1984@gmail.com
www.lotuspress.co.in

Vedic Mathematics New Horizons AdvanceLessons

ISBN: 978-81-8382-298-5

Printed & Published by : **Lotus Press Publisher & Distributors,** New Delhi-02

FOREWORD

Swami Bharti Krishna Tirtha ji Maharaj well demonstrated potentialities of Ganita Sutras. The illustrative applications of Ganita Sutras worked out by Swami ji, attracted the attention of the world of mathematics. But the Discipline again has started having signs of stagnation for want of further clues. As it appears that the scholars are getting confined, to the domain of the illustrative demonstrations given by Swami ji.

Being clueless, as to further applications of Ganita Sutras, the subject has not entered the main stream of knowledge. Now with these two volumes of initial and advance lessons of Dr. S. K. Kapoor, new hope of revival of the Ancient Discipline has become lively. Inspired by the works of Dr. Kapoor, we, the i360 Staffing and Training Solutions Pvt. Ltd., company in collaboration with the Unicon Trust and Paramount Knowledge Pvt. Ltd. Company are planning to take up the Project: 'Vedic Mathematics' for its dissemination and research.

We shall be requesting Dr. S. K. Kapoor to help us structure Vedic Mathematics Teachers Courses and soon we would like to take up these courses in the right earnest.

I am sure that these LESSONS, as the Title conveys shall be leading the Students, Teachers, Scholars and Institutions dedicate for 'Dissemination and Research' in the field of Vedic Mathematics, to new horizons of the Discipline.

(Mrs. Sonia Nagpal)

Chairperson
i360 Staffing and Training
Solutions Pvt. Ltd.
New Delhi

PREFACE

Vedic Mathematics, Science and Technology is the ancient Discipline which for some historic reasons remained dormant for some centuries.

Credit goes to Swami Krshna Tirtha Ji Maharaj to focus the attention of present generation about the potentialities of Ganita Sutras. Though the Ganita Sutras text consist of sixteen Sutras and 13 Upsutras together availing only 520 letters and as such is a very short text but the same, in the words of Swamiji, these cover whole range of mathematics known to us and even beyond that of the Vedic order.

These two volummes of Vedic Mathematics lessons, volumme-1 initial lessons (70) and volumme-2 Advance lessons (78) aim to give insight about the potentialities range of these Sutras.

Present lessons are in continuity of the 'Initial Lessons', and in that sequence, these are being designated as 'Advance Lessons'.

These lessons are grouped as of general, specific and special features and as 'Ancient Wisdom' and Knowledge systems of Ganita Sutras. Text of Ganita Sutras and Upsutras is being added as Annexure-1 and 2.

For conceptual help, it would be helpful to refer to the initial lessons.

Dr. S.K. Kapoor

CONTENTS

SECTION – 2: SPECIFIC FEATURES

SECTION – 3: SPECIAL FEATURES

SECTION – 4: KNOWLEDGE SYSTEMS OF GANITA SUTRAS

SECTION – 1: GENERAL FEATURES

1.

0	1	2	3	4	5	6

Hyper cubes 0, 1, 2, 3, 4, 5, 6

2. Measure and Measuring rod
3. As to the existence of 4-space
4. 3-Space Body to 4-Space Body
5. Numbers Line to Square Number Line
6. Bend at the Middle of Squares Numbers Lines
7. Interval, Square and Cube, as Hyper Cubes 1, 2 and 3
8. $(A+2)^N$, N=1, 2, 3
9. Transition From Linear Order to Spatial Order
10. Let us have a Fresh Look at the Cube
11. Geometric Envelope of Cube
12. 26 Basic Elements
13. Vedic Alphabet Format Features
14. Mathematics of '2 as 1' and '1 as 2'
15. Geometry NVF (108)
16. Vedic Mathematics Answer

1

0	1	2	3	4	5	6

Hyper cubes 0, 1, 2, 3, 4, 5, 6

1. Vedic knowledge has pure as well as applied values.
2. Vedic Mathematics, Science and Technology is the Ancient wholesome Vedic Discipline.
3. Vedic mathematics is at the core of the Discipline of 'Vedic Mathematics, Science and Technology'.
4. The establish Vedic processing system is of two fold features, namely, *Sankhiya Nishtha* and *Yoga Nishtha*.
5. Both these folds run parallel to each other and are complementary and supplementary of each other at every phase and stage of processing.
6. *Sankhiya Nishtha* ultimately avails artifices of numbers along geometric formats and on the other hand, the *Yoga Nishtha* avails dimensional frames parallel to artifices of numbers.
7. This processing feature in terms of artifices of numbers presumes the existence of dimensional frames and on the other hand the processing along dimensional frames would presume the existence of artifices of numbers.

8. Each dimensional frame, accordingly is parallel to distinct place value system of the artifices of numbers.
9. Of these, parallel to ten place value system of numbers is the 5-space format.
10. For the present, to begin with, the present processing pursuit is of the features of a ten place value system of 5-space format.
11. And, the Ganita Sutras as complete scripture while accepting sole syllable *Om* (ॐ) as the transcendental source reservoir (5-space), avail sixth vowel (ए) as the first letter (of the first Ganita Sutras itself), and as such the mathematics (and, Vedic Mathematics, Science and Technology discipline as well) is to be worked beginning with the 'features' of artifices 6 / 6-space.
12. This beginning, as such, is to be the beginning along the format of hyper cube 6, the representative regular body of 6-space.
13. The coverage range of 6-space domain/hyper cube 6, accord-ingly is to be in terms of point, interval, square, cube and hyper cubes 4, 5 and 6, as representative regular bodies of 0 to 6-space.
14. This range of hyper cubes 0 to 6, as such is the range and format of *Sathapatya* measuring rod.
15. The *Sathapatya* measuring rod accept Lord Vishnu, the Lord of 6-space/Sun/Atman/*Vishnu Lok*/*Go lok*/*Pursha* format presiding deity and Lord Brahma, the overlord of Creator space (4-space), as the presiding deity of the measure of the *Sathapatya* measuring rod.
16. As such, for such beginning of processing in terms of the *Sathapatya* measuring rod, the first lesson, as such is to be to acquaint oneself with the point, interval, square, cube, hyper cubes 4, 5 and 6, as the representative

regular bodies of 0-space to 6-space.

17. For it, it would be convenient, to have respective symbols for these bodies, and, the one set of symbols being availed here are as captioned, at the outset of this lesson.
18. Unless and until, the context is otherwise, these very symbol may be taken as the symbols of corresponding spaces as well as of the representative regular bodies of respective spaces.

◆ ◆ ◆

2

MEASURE AND MEASURING RODS

1. We are well acquainted that in 3-space, we measure along a line (straight line) to compute length, area and volume as a single axis measure, pair of axes measure and as triple axes measure.
2. This is a process of approaching 3-space domain (solids), in three steps as length, length and breadth and as length, breadth and height.
3. It is like approaching cube in terms of interval, square and cube within a cube.
4. Conceptually it means approaching domain (here 3-space domain) or in general a space (here 3-space) in terms of dimension, which in the context of 3-space, comes to be of the format and features of a linear axes which in other words means as that 1-space playing the role of dimension of 3-space.
5. So much so, is alright within the present mathematical tools in the form of a measuring rod of a linear measure.

6. However, if we pose within the modern mathematical systems, what would be the answer to the question as that, what is the dimension of 1-space?
7. And in general, the poser can be as to what is the dimension of n-space?
8. Vedic mathematical answer to it is, as that (n–2) space plays the role of dimension of n space.
9. *Sathapatya* Upved of *Atharavved*, which is the source reservoir of Vedic mathematical system (and in general, of Vedic mathematics, science and technology systems), in its scripture designated '*Mansara*', which means 'essence of measure' enlightens us as that Lord Vishnu, the overlord of 6-space, is the presiding deity of the measuring rod and Lord Brahma, creator the supreme, the overlord of 4-space, is the presiding deity of the measure of this measuring rod (of *Sathapatya Upved*)
10. Let us have a pause here and permit the transcending mind to think, mediate and transcend through this enlightenment and to glimpse its features and to be face to face with the wisdom as that (4-space as measure) plays the role of dimension of 6-space (as domain/ measuring rod).
11. The ancient wisdom of other Vedic scriptures, specifically provide as that the general difference between domain and dimension is '2', which leads to the conclusion as that the n-space, n–2-space plays the role of its dimension.
12. Working this out shall be taking us to (–1) space playing the role of dimension of +1-space, 1-space playing the role of dimension of 3-space, 2-space playing the role of dimension of 4-space and so on.
13. With it 1-space emerges to be of features within a single dimension frame with –1-space playing this role of dimension.

14. 2-space emerges as of features of pair of dimensions frame with 0-space playing this role of dimension.
15. Further 3-space emerges to be of features of 3 dimensions frame with 1-space playing this role of dimension.
16. Ahead 4-space would emerge as to be of features of 4 dimensions frame with 2-space playing this role of dimension.
17. Ahead, 5-space is to be of features of 5 dimensions frame with 3-space playing this role of dimension and 6-space shall be of features of 6-space dimension frame with 4-space playing this role of dimension.

◆ ◆ ◆

3

AS TO EXISTENCE OF 4-SPACE

1. One way to comprehend and to conceptualize about the reality of existence of 4-space would be to have a fresh look at our well known geometric bodies namely 'point, interval (line), square (surface) and cube (solid)'.
2. Let us have a fresh look at the set-up of interval (line) and to comprehend and to have an insight of this set-up as that it is a format/path of a moving point.
3. Further have a fresh look at the set-up of the square (surface) and to comprehend and to have an insight of this set-up as a format/path of a moving interval/ line.
4. Still further to have a fresh look at this set-up of cube (solid) and to comprehend and to have an insight about this set-up as a format/path of a moving square/ surface.
5. Let us have a pause here and to permit the transcending mind to comprehend these features of interval, square and cube as formats/paths of moving point, interval and square respectively.

6. It shall be bringing us face to face with these features as that interval as body of 1-space is of the format/path of a moving point body of 0-space.
7. Further as that square as body of 2-space is a format/path of a moving interval, the body of 1-space.
8. And ahead as that cube as body of 3-space is a format/path of a moving square, the body of 2-space.
9. Let us further have a pause here and to have a fresh look at the above features as that 0-space body leads to 1-space format, 1-space body leads to 2-space format and 2-space body leads to 3-space format.
10. And as such, the natural conclusion would be that 3-space moving body shall be leading to 4-space format.
11. One shall sit comfortably and permit the transcending mind to comprehend and imbibe the features and values of 4-space format.

◆ ◆ ◆

4

3-SPACE BODY TO 4-SPACE BODY

1. To reach at representative regular body of 4-space as fourth member of the sequence of representative regular bodies of dimensional spaces, one shall have a fresh look at the set-ups of interval, square and cube being the first three members of the sequence.
2. Let us have a fresh look at the set-up of interval. It is a set-up with a pair of end points.
3. Taking end points as 0-space bodies and the set-up between these pair of end points as a set-up of 1-space content, it shall be having an expression (A^0, A^1, A^0).
4. This also would admit expression as $2A^0, A^1$.
5. This expression $(2A^1)$ conceptually is a domain (A^1) within boundary $(2A^0)$.
6. This as such, may be accepted as a formulation domain is to boundary :: A^1: $2A^0$.
7. Now the square be looked afresh and it would lead to a set-up within four boundary lines leading to formulation :: A^2: $4A^1$.
8. Likewise the set-up of the cube shall be leading to the formulation :: A^3: $6A^2$.

9. Let us have a pause here and permit the transcending mind to transcend through these formulations for the set-ups of interval, square and cube and these shall be bringing face to face with a single formulation domain : Boundary :: A^n: $2nA^{n-1}$, n=1, 2, 3.
10. For n = 4, it shall be leading us to formulation for 4-space representative regular body namely for hyper cube 4 as domain : Boundary :: A^4: $8A^3$.
11. Let us have a pause here and think, meditate and transcend through the features of above formulations of hyper cube 4 which shall be bringing us face to face with hyper cube 4 as being enveloped within solid boundary of eight components.
12. One may further have a pause here and to permit the transcending mind to transcend through the origin of three dimensional frame and to glimpse 3-space splitting into eight octants and parallel to it cube splitting into eight sub cubes with corner points of these eight sub cubes/eight sub octants having placement at the seat of origin itself.
13. This as such shall be bringing us face to face an additional feature as that the location of hyper cube 4/4-space is at the centre of cube/origin of 3-space.
14. As such the domain of hyper cube 4 with solid boundary of eight components, as fourth member of the sequence of representative regular bodies of dimensional spaces, shall be accepting symbolic representation, in continuity of interval, square and cube as follows:

1	2	3	4
			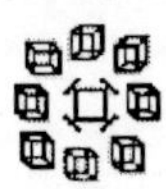
Interval	Square	Cube	Hyper cube 4
A^1: $2A^0$	A^2: $4A^1$	A^3: $6A^2$	A^4: $8A^3$

◆ ◆ ◆

5

NUMBERS LINE TO SQUARE NUMBER LINE

1. Transition from linear order 3-space to spatial order 4-space shall be having a transition from linear measure to spatial measure.
2. Ganita Sutra-1 *Ekadhiken Purvena* rule leads to numbers line

3. Ganita Sutra-7 संकलव्यवकलनाभ्याम् addition and minus rule leads to the square numbers lines as follows:

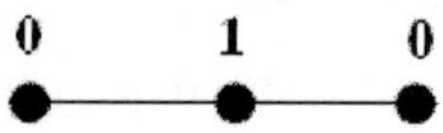

$0 + 1 + 0 = 1 = 1^2$

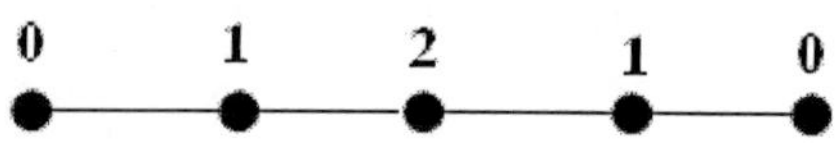

$0 + 1 + 2 + 1 + 0 = 4 = 2^2$

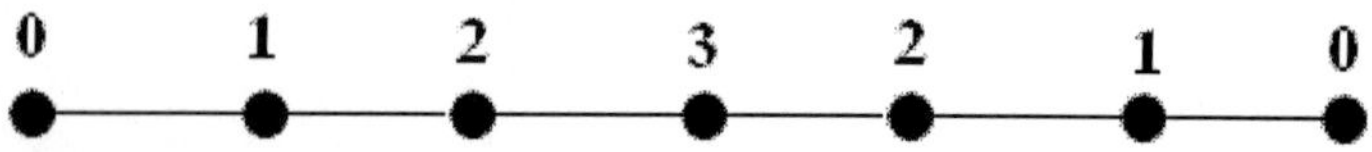

$0 + 1 + 2 + 3 + 2 + 1 + 0 = 9 = 3^2$

$$0 + 1 + 2 + 3 + \text{-----} N + \text{------} + 3 + 2 + 1 + 0 = N^2$$

4. One may have a pause here and have a fresh look at the above set-ups of squares lines, which have a pair of streams flow from the middle towards either end in descending orders.
5. Looking from another angle, these set-ups would be of the features of a pair of flow sequences from either end to the middle.
6. It as such shall be making seat of the middle of such line as to be of a spatial order.
7. It is this transition and transformation from the linear seat for the middle for a line into a spatial features for the seat of the middle which deserves to be comprehended and to be imbibed.
8. It is this comprehension which shall be providing desired insight into the set-ups within spatial order 4-space.

◆ ◆ ◆

6

BEND AT THE MIDDLE OF SQUARES NUMBERS LINES

1. The squares numbers line permit bend at the middle as follows:

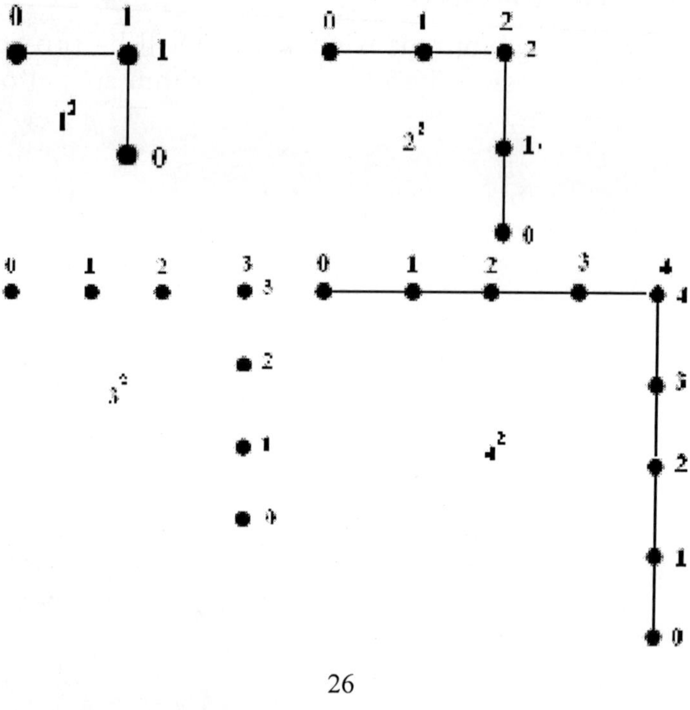

2. It is this feature of artifices along geometric formats which deserves to be comprehended and imbibed to have transition from linear measures to spatial measures.

3. These formats shall be leading to grid formats/matrix values format 1×1, 2×2, 3 3, 4×4 and so on.

4. Amongst other features, the one prominent feature of these grid formats is that these can be smoothly availed for organization of cubic values as follows:

   1 2 = 1 + 2 = 3 $3 + 5 = 8 = 2^3$
   2 3 = 2 + 3 = 5

   1 2 3 = 1 + 2 + 3 = 6 $6 + 9 + 12 = 27 = 3^3$
   2 3 4 = 2 + 3 + 4 = 9
   3 4 5 = 3 + 4 + 5 = 12

   1 2 3 4 = 1 + 2 + 3 + 4 = 10 $10 + 14 + 18 + 22 = 64 = 4^3$
   2 3 4 5 = 2 + 3 + 4 + 5 = 14
   3 4 5 6 = 3 + 4 + 5 + 6 = 18
   4 5 6 7 = 4 + 5 + 6 + 7 = 22

   And so on.

5. There are other features like that of reflection operation, which shall be splitting square into 4 sub squares, and this together with the pair of surfaces, shall be leading to availability of eight cubes at a time, which shall be taking to set-up of 3-space as a set-up of eight octants, which shall be providing transition from spatial measures to solid measure.

◆ ◆ ◆

7

INTERVAL, SQUARE AND CUBE AS HYPER CUBES 1, 2 & 3

1.

1	2	3
		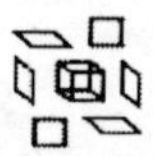
Interval	Square	Cube

2. Let us have a fresh look at the set-up of interval (H). It has a pair of end points (bodies of 0-space).
3. The set-up in between the pair of end points, being designated as length part of the interval, as such is a content lump of 1-space, and so to be taken as 1-space body.
4. Further as, (–1) space plays the role of dimension of (+1) space, as such it plays its stitching role for 1-space content lump.
5. Still further as the middle of this set-up permit approach from its either side, which that way provides a seat, to be designated as square origin source.
6. These four distinct features (of 0-space, –1-space, 0-space, +1-space and 2-space) together make the set-up of interval as four fold set-up.

7. This four fold set-up, as of four consecutive dimensional content lumps, synthesized as such, as interval makes it of the format of four fold measuring rod of creator space (4-space).
8. Interval, this way, as of features, set-up and synthesis of four fold dimensional contents of four consecutive dimensional spaces (here –1-space content lump, 0-space content lump, +1-space content lump and 2-space content lump) is designated as hyper cube 1.
9. The four folds of hyper cube 1 are distinctively designated as dimension fold, boundary fold, domain fold and origin fold.
10. With it (–1) space plays the role of dimension fold of hyper cube 1, 0-space plays the role of boundary fold of hyper cube 1, 1-space itself plays the role of domain fold of hyper cube 1 and ahead, 2-space plays the role of origin fold of hyper cube 1.
11. This quadruple set of values would permit expression in terms of artifices values as (–1, 0, 1, 2).
12. This way quadruple (–1, 0, 1, 2) is designated as four fold manifestation values of hyper cube 1.
13. Accordingly hyper cube 1 is accepted as a manifestation layer of four folds (–1, 0, 1, 2)/(–1-space, 0-space, 1-space, 2-space).
14. One may have a pause here and take note as that while interval is taken as hyper cube 1, this, that way amounts to giving a focus to the domain fold of the manifestation layer.
15. However, this focus can be upon either of the four folds and that, that way would have appropriate way to approach hyper cube 1, and like that, in general to any of the hyper cubes.
16. This focus for interval as hyper cube 1, shall be helping us focus upon the domain fold features of 1-space

content lump. But this, in no way shall mean that only domain fold is to be taken for its role to know the properties of 1-space.

17. 1-space, and like that every other space, is to play different roles, of these four prominent roles of manifestation layer format of creator space (4-space) are (1 as dimension fold, 2 as boundary fold, 3 as domain fold and 4 as origin fold).
18. This four fold role for 1-space, as such would lead to 4 4 matrix format of order and values as such:

–2	–1	0	1
–1	0	1	2
0	1	2	3
1	2	3	4

19. One may have a pause here and permit the transcending mind to comprehend and imbibe the values of four manifestation layer, namely (*i*) –2, –1, 0, 1, (*ii*) 1, 0, 1, 2, (*iii*) 0, 1, 2, 3 (*iv*) 1, 2, 3, 4.
20. It would be a blissful exercise to approach square and cube as hyper cubes 2 and 3 respectively and to reach at their 4 4 matrix format as under:

–1	0	1	2
0	1	2	3
1	2	3	4
2	3	4	5

And

0	1	2	3
1	2	3	4
2	3	4	5
3	4	5	6.

◆ ◆ ◆

8

$(A+2)^N$, N = 1, 2, 3

1. Let us again have a fresh look at the set-ups of interval, square and cube with a focus upon the geometric components together constituting these set-ups as follows:

1	2	3

Interval	Square	Cube

2. Let us have look at the set-up of interval. It consists of three components namely pair of end points and length.
3. The square is a set-up of four corner points, four boundary lines and surface area.
4. The cube is a set-up of eight corner points, 12 edges, 6 surface plates and space volume.
5. The geometric components of interval can be expressed as $A^1 + 2\,A^0 = (A + 2)^1$

6. The geometric components of square can be expressed as $A^2 + 4A^1 + 4A^0 = (A + 2)^2$
7. The geometric components of cube can be expressed as $A^3 + 6A^2 + 12A^1 + 8A^0 = (A + 2)^3$
8. These formulations for the expressions of the geometric components of the set-ups of interval, square and cube, together shall be accepting a single formulation = $(A + 2)^N$, N = 1, 2, 3.
9. One may have a pause here and think, meditate and transcend through the features of the formulation (A + 2) and see that this reach from A to A + 2 is of the reach from dimension (here A) to domain (here A + 2).
10. Further have a fresh look at the formulation $(A + 2)^N$ N =1, 2, 3 and to comprehend and imbibe these features as that of single axis, pair of axes and triple axes, parallel to the set-up of a three dimensional frame.
11. It as such shall be focusing upon the linear order set-ups reaching at domain.
12. Now let us chase the formulation $(A - 2)^N$, N = 1, 2, 3.
13. $(A - 2)^1$, shall be leading us to A – 2 which in the context of set-up of the interval would mean that when both end points are stripped off, we would be left with the length part only.
14. This, as such would mean that when we shall be transiting from $(A + 2)^1$ to $(A - 2)^1$, we shall be transiting from closed interval to open interval.
15. It is this transition which shall be helping us have an insight as to the transition from the mathematics of 1-space in the role of dimension to -1-space in the role of dimension.
16. Further It would be relevant to note that as (–1) space plays the role of dimension of (+1) space and ahead (–1) space plays the role of dimension of 3-space, it will make

that (–1) space plays the role of dimension of dimension of 3-space.

17. In general it would amount to reaching from dimension of dimension to dimension to domain (domain fold) of the manifestation layer.
18. Now $(A - 2)^2 = A^2 - 4A + 4$, in the context of square shall be leading to the format of a surface (of square) having its four corner points being intact but its all the four boundary lines having been stripped off.
19. This, this way shall be giving us insight about the pair of dimensions role at the dimension of dimension level, and consequential differences of the mathematics at that level from that of the mathematics of pair of dimensions at the dimensions level.
20. Further $(A - 2)^3 = A^3 - 6A^2 + 12\ A^1 - 8$ when viewed in the context of the format of the cube it shall be leading us to the set-up of the cube as a volume along with 12 edges while its all the six surface plates together with all the eight corner points standing stripped off.
21. One shall have a pause and permit the transcending mind to remain in prolonged sittings of trans to be face to face with the mathematics at dimension of dimension level of the domain folds of the manifestation layers in the context of linear order of 3-space.

◆ ◆ ◆

9

TRANSITION FROM LINEAR ORDER TO SPATIAL ORDER

1. Ancient wisdom enlightenment for transition from linear order to spatial order is of the quadruple values (1, 2, 3, 8).
2. Let us have a fresh look at this quadruple values set-ups (1, 2, 3, 8).
3. These quadruple values (1, 2, 3, 8) permit re-organization as ($1 \times 1, 1 \times 2, 1 \times 3, 2 \times 4$).
4. Let us have a fresh look at these re-organized values ($1 \times 1, 1 \times 2, 1 \times 3, 2 \times 4$)
5. First features of this re-organization is that it admits split as two parts namely ($1 \times 1, 1 \times 2, 1 \times 3$) and ($2 \times 4$).
6. The first part, namely ($1 \times 1, 1 \times 2, 1 \times 3$) is of rectangular set-ups while the second part (2×4) as ($2 \times 2 \times 2$) is of a cubic format features.
7. This, this way would mean to have surfaces and solid approach.
8. The formulation (एक), the first formulation of the text of Ganita Sutra-1, is a set-up of pair of letters i. (ए) ii. (क्), which are of artifices values i. (6) ii. (4) respectively.

9. The formulation (एक) as artifice pair (6, 4) that way envelope the gap as artifice value (5).
10. Now let us have a look at the artifices pair (6, 4) which admits re-organization as [(3 + 3), (2 + 2)].
11. As such the middle value (5), which admit re-organization as (3 + 2), would help us comprehend that the mathematics of the gap, as such is to avail artifices 3 and 2 and parallel to it (3-space) and (2-space) which would mean 'solid and surface' formats.
12. Other feature of the pair (6, 4) is that 6 + 4 =10 is double of the middle value (5).
13. It as such would help us comprehend and imbibe the value as that that the middle is to be worked out in terms of a half unit.
14. It is this feature which would help us comprehend and imbibe the features of the quadruple (1, 2, 3, 8) permitting re-organization as (1 × 1, 1 × 2, 1 × 3, 2 × 4) which is for the first half is having multiple (1) but for the second part it is having the multiple (2).
15. It is this sequential order of transition from linear order to spatial order, which in ascending order gives us the multiples as 1 and 2 in that sequence and order.
16. However, when the sequence is to be of other orientation for the squares number line (of order 0, 1, 0 = 0 + 1 + 0 = 1 = 1 × 1), which would mean reaching the middle from the other end, it shall be taking us from '2' to '1'.
17. This reach from '2' to '1' is a reach from '1' to ' ½ '.
18. Further, this reach from 1 to 2 and from 1 to ½ with reverse orientation, that is from ½ to 1 is not going to be of the same tracks, formats and values.
19. To comprehend this distinguishing feature of the pair of tracks would require us to take note as that the mathematics and domain level and at dimension level is of different features.

20. The difference is mainly there as that pair of dimensions for their synthesis shall be requiring a stitch/glue of the value of the unit of dimension of dimension.
21. It in the context of 1-space as dimension shall be taking us to (–1) space being the dimension of dimension.
22. Accordingly the synthesis of pair of linear dimension of values 1 and 1 together though would be of domain value (of 1-space) being 1 + 1 = 2 but out of it the synthesis stitch/glue value is to be compensated.
23. As the unit synthetic stitch/glue value being supplied by (–1) space as dimension of dimension, so this unit value (–1), when would be compensated out of the 1 + 1 = 2 value it shall be leading to 2 – (–1) = 3.
24. Accordingly the track approach at domain measure for 1-space as dimensional order shall be leading to 1 + 1 = 2 but the same in the reverse orientation shall be leading us to the value 2 – (–1) = 3.
25. It is this feature that to reach the gap from its both ends would mean to have a coverage in terms of artifices values (2, 3).
26. One shall have a pause and permit the transcending mind to comprehend this feature and to imbibe its values to be face to face with the phenomenon of transition from the linear order set-up of 3-space with cube as its representative regular body having 26 geometric components envelope for its space volume to spatial order 4-space with hyper cube 4 as its representative regular body, the same is to be covered accepting it to be the phenomenon of quadruple values (1, 2, 3, 8).

◆ ◆ ◆

10

LET US HAVE A FRESH LOOK AT THE CUBE

1. To have a fresh look at cube () to evaluate as to how much you know about the cube ().

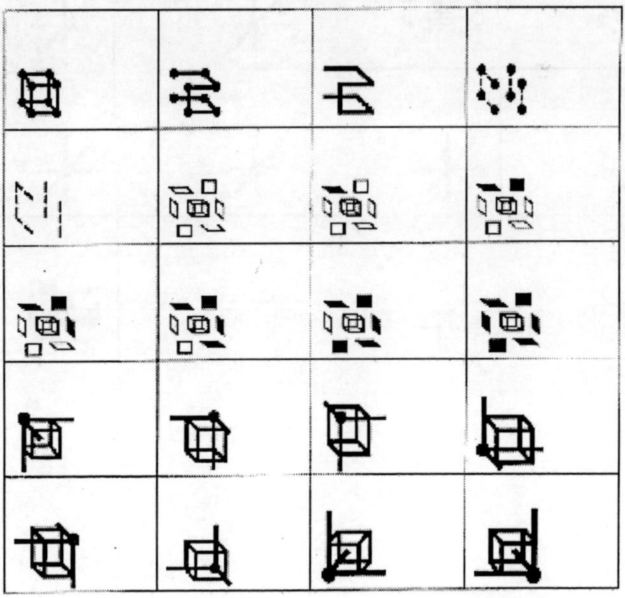

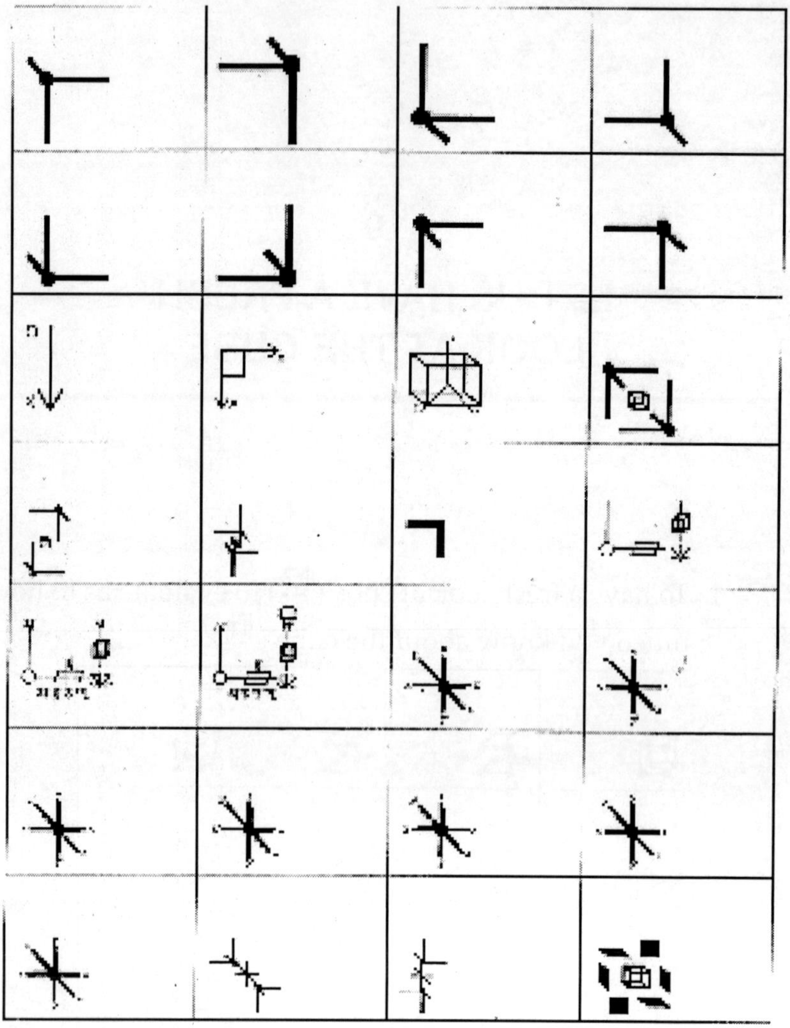

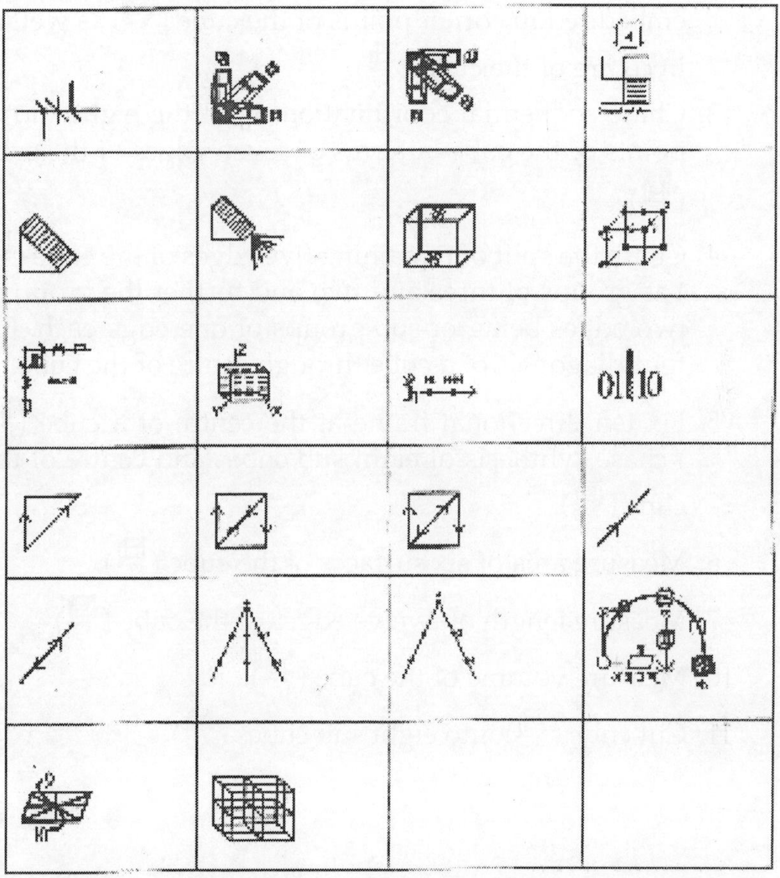

2. Have a look at the following depictions about different structural features of the cube and evaluate your comprehension of them.
3. Enlist the different structural component of the cube () and to have a fresh look at each of its structural components to evaluate as to how much you know about them (that is about, corner points (●), edges (–), surfaces (■) , volume, diagonals, dimensional frame, dimensions, origin, its eight sub cubes () split and all that).
2. Further, to have a fresh look at three dimensional frames

embedded in corner points of the cube ( ) as well as in centre of the cube ( ).

3. Chase sequential coordination of all the eight corner points of the cube in terms of seven edges of the cube ( ).

4. Chase the split of remaining five edges of the cube ( ) as groups of three and two and further the group of two edges being of sub groups of one edge each. Fix four diagonals of a cube through centre of the cube.

5. Fix ten directional frame at the centre of a cube ( ) chase synthesis of eight sub cubes and centre of the cube ( ).

6. Measure area of six surfaces of the cube ( ).

7. Measure length of twelve edges of the cube ( ).

10. Measure volume of the cube ( ).

11. Cut cube ( ) into eight sub cubes ( )

◆ ◆ ◆

11

GEOMETRIC ENVELOPE OF CUBE

1. Geometric envelope of a cube consist of eight corner points, twelve edges and 6 surfaces.
2. Geometric components formulation $(A + 2)^3 = A^3 + 6 A^2 + 12 A^1 + 8 A^0$, has two parts, firstly A^3 and secondly $6 A^2 + 12 A^1 + 8 A^0$.
3. One feature of this split is that it is of '1' and '26' components.
4. The first part value '1' is a unit cube value as $1^3 = 1$.
5. The combined value of first part and second part is $1 + 26 = 27 = 3^3$.
6. This, as such shall be leading us to as that cube without its envelope is 1^3 and that the cube along with its envelope is of value 3^3.
7. This, this way would also help us comprehend and to have an insight as that in the process there is a jump over 2^3.
8. One shall have a pause and permit the transcending mind to comprehend this feature of jump from 1^3 to 3^3 as a jump over the sequential gap value 2^3.

9. One shall further permit the transcending mind to transcend through the triple value ‘1^3, 2^3, 3^3.
10. For it one shall sequentially transcend through (*i*). (1, 2, 3) (*ii*). (1^2, 2^2, 3^2) and (*iii*). (1^3, 2^3, 3^3).
11. One may again have a pause here and to take note that, one, two and three is the only triple which is unique in many ways, and even it can be taken as consecutive three primes.
12. This as such with ‘2’, which is the only even prime, at the middle of the triple, deserves to be chased very gently and ultimately one would find that all the knots and mathematical problems which shall be surfacing while approaching the set of artifices of numbers and corresponding dimensional frames, the same would be there only because of the linear order approach not taking care of the need for transition from linear order to spatial order.
13. It is the geometric envelope (of 26 components and accordingly of artifice value 26) which is to help attain needed transition from linear order to the spatial order.
14. Ancient wisdom approaches the existence phenomenon of the whole range from Earth to Sun as a six steps long range of five basic elements and sixth as Sun, in terms of 26 elements as affine values of artifices 1 to 26 each to be of four quarters set-up and thereby there being the attainment of order of the values 26 4 = 104 as a creator space (4-space) as much as that 4 = 1 +1 + 1 + 1, ahead as sequential progression 1 + 2 + 3 + 4 = 10 makes the whole processing features to be availed for such coverage of Earth to Sun range.

◆ ◆ ◆

12

26 BASIC ELEMENTS

1. Ancient wisdom accepts 26 basic elements.
2. These 26 basic elements are the affine values of artifices 1 to 26.
3. Parallel to it there are 26 metres (*chandus*)/measures/values /virtues.
4. Within 4-space/hyper cube 4, static 12 edged cube becomes dynamic 13 edged cube with its 13th edge being the additional edge acquired by the static cube along its dynamic flow path within 4-space/hyper cube 4 domain.
5. There being spatial order of 4-space, as such it becomes of $13 \times 2 = 26$ artifices value.
6. Further as hyper cube 4 is enveloped within solid boundary of eight components, as such it at the boundary yields $13 \times 8 = 104$ coordinates accepting fixation as 4 quarters values for 26 values of the spatial order fixation for the 13 edged cube within hyper cube 4 domain.
7. These 26 values fixation for the 13 edged cube within

hyper cube 4 domain manifest 26 components geometric envelope for the cube.

8. These 26 artifices values permit re-organization as $5\times5 + 1\times1$.
9. It shall be that way splitting 26 elements range as that of 25 elements range and ahead single element range.
10. This split also focus upon 5×5 matrix format, and that way upon artifice '5'.
11. It this way brings to focus the six steps long range of five basic elements designated as Earth, Water, Fire, Air, Space and Ahead the sixth being the Sun.
12. There are precisely 26 primes (including 1) up till the range of artifices 1 to 100, that is from 1×1 to 10×10.
13. One shall have a pause and permit the transcending mind to remain in prolonged sittings of trans to transcend through the ancient enlightenment based wisdom and knowledge preserved in lively traditions, as that (*i*). all that the human mind will reach at the values of the existence phenomenon shall be found lively within the Vedas (*ii*). All human articulated languages shall be the adaptations of the Vedic alphabet format.
14. One may again have a pause and permit the transcending mind

◆ ◆ ◆

13

VEDIC ALPHABET FORMAT FEATURES

1. Vedic alphabet format features are parallel to affine values of artifices 1 to 26 accepting pairing their of as a range of 52 letters accepting organization as range of 50 letters in between the pair of end letters.
2. It is the coverage of the range in between the pair of end points, that is the middle range of 50 letters, which is to remain under focus.
3. This pairing value together with 26 affine artifices values, which is to be of ultimate operative phases and stages. Of these 5 × 5 = 25 consonants range, as half of the middle range, which as such is to be the ultimate operative values, which provide format for 5 ×5 *Varga* consonants. Accordingly there are 25 *Sankhiya* elements.
4. It is in terms of these 25 artifices permitting organization as 5× 5 format that the artifice '5' gets prominence and the same, as such becomes the 5 steps long range of 5 basic elements designated as *Panch Mahabhut*, titled, Earth, Water, Fire, Air and Space.

5. One may have a pause here and permit the transcending mind to remain in prolonged sittings of trans to comprehend and imbibe the values of affine set of artifices 1 to 26, as alphabet letters values and in terms of it glimpse the way 26 letters popular English letter alphabet is availing the features, values and virtues of Vedic alphabet format.
6. It would be a blissful chase for the sadkhas to glimpse the way Vedic alphabet format feature would help bring to focus the basic features, values, virtues and order of classical and orthodox English Language vocabulary by associating number values format for letters A to Z as 1 to 26 artifices in that sequence and order.
7. This being the affine values association of artifices 1 to 26, as such these would parallel to 26 elements, shall be accepting pairing their of, and as such the English language, in this light, is to be accepted as a 'pairing' language.
8. One may again have a pause here and comprehend these pairing features of affine values of artifices as AA = 1 + 1 = 2 and ZZ = 26 + 26 = 52. Like that the WORLD, as English vocabulary word, as it avails the letters W of number value format (in short NVF (W) = 23 and NVF (O) =15, NVF (R) = 8 and NVF (D) = 4 with their summation value 23 + 15 + 18 + 4 = 60 = NVF (WORD).
9. The pairing (summation value) for ENGLISH, shall be NVF (ENGLISH) = 74 = NVF (PAIRING).
10. Chase of NEW TESTAMENT (BIBLE) shall be taking us to NVF (RANGE) 1 to 242 and this, as such shall be giving us bliss as that we have to cover only a short range of number value format for enlightenment as

that NVF (GOD) = 26.

11. This chase of classical and orthodox English vocabulary of 26 letters alphabet in terms of features of Vedic alphabet format as per affine values of artifices range 1 to 26 would be blissful as it shall be giving us confidence and satisfaction about the Vedic traditions lively claims.

11. NVF (WORD) = 60 = NVF (FOUR) shall be further giving us confidence as that it is all about 4-space of spatial order. Further as that NVF (FOUR) = 60 = 2 + 58 = NVF (2 TWO).

◆ ◆ ◆

14

MATHEMATICS OF '2 AS 1' AND '1 AS 2'

1. For transition from linear order of 3-space to spatial order of 4-space, transition need be had for mathematic of '1' to mathematic of '2 as 1' and '1 as 2'.
2. This shall be a transition for the format of triple (0, 1, 0) to that of the format of triple (1, 0, 1).
3. The geometric expressions for these triples shall be:

Triple	Geometric expression
(0, 1, 0)	A^0 A^1 A^0 0 1 0
(1, 0, 1)	A^1 A^0 A^1 1 0 1

4. The triple (1, 0, 1) with above corresponding geometry format shall be leading to a set-up for interval within spatial order as having been fixed at its middle as a pair

of intervals, synthesized as an interval with synthetic joint at the middle in terms of dimension of dimension unit of spatial order 4-space.

5. One may have a pause here and permit the transcending mind to remain in prolonged sittings of trans to comprehend and imbibe the features, order, values and virtues of this synthetic set-up of the interval to be face to face with the number values formats of affine artifices values 1 to 26 as 26 letters alphabet leading to : d
NVF (INTERVAL) = 9 + 14 + 20 + 5 + 18 + 22 + 1 + 12 = 101.

6. The sequential chase of the artifices values (9 + 14 + 20 + 5 + 18 + 22 + 1 + 12 = 101 as the spatial order is to lead to the manifestation layer (2, 3, 4, 5) with 2-space in the role of dimension and 5-space in the role of origin, as such the hyper cube 4 format with 2-space as dimension, 3-space as boundary, 4-space as domain and 5-space as origin shall be the format on which the formulation 'Interval' as sequential range of artifices values '9 + 14 + 20 + 5 + 18 + 22 + 1 + 12 = 101' are to be chased.

7. The geometric expression for this format is to be taken as follows:

8. The sequential chase with first step value as '9', shall be taking us to 9 geometries range of 4-space/9 versions of hyper cube 4.

9. Here, it would be relevant to note that artifice 5 is at the middle of this range of artifices 1 to 9.

10. The second step value of artifice 14, as such leads us to the four fold manifestation layer (2, 3, 4, 5) with (2 + 3 + 4 + 5) = 14, as expression for hyper cube 4

11. Here, it would be relevant to note that 9 + 5 = 14, as a step ahead shall be leading us to 14 + 6 = 20.

12. Third step artifice value T = 20 admits re-organization as $20 = 4 \times 5$.

13. It is this organization which shall be leading us to the features of 4-space playing the role of dimension of 5-space and boundary having fixation in terms of $4 \times 5 = 20$ coordinates of solid dimensional order of 5-space.

12. Further It would be relevant to note that artifice 20 with its organization 9 + 7 + 3 + 1 shall be a *Divya Ganga* flow format for 2 fold transcendental reach for the middle (which here is a transcendental domain of artifice 5/5-space) from its both end of 1-space leading to 3-space and ahead there being 5-space. From the other end decadence from 9-space to 7-space as a dimensional order, ahead shall be leading to 5-space as dimension of dimension of 9-space.

13. With it, one may comprehend and imbibe the values of transcendental domain of artifice 5/5-space as artifice value of the fourth step of the formulation interval.

14. Step 5 which avails the artifices value R = 18 accepts re-organization as four fold manifestation layer (3, 4, 5, 6) with 3 + 4 + 5 + 6 = 18.

15. A step ahead at as a step 6 of artifice value V = 22, is of re-organization value of four fold manifestation layer (4, 5, 6, 7) with 4 + 5 + 6 + 7, the manifestation layer of hyper cube 6.

16. Step 7 avails artifice A = 1.

17. Here, It would be relevant to note that with 5-space as origin fold of 4-space, shall be, at this phase and stage with role of 5-space as dimensional order of 7-space, shall be re-setting the measuring process, and as such it is to be of value '1'.

18. Here, It would be relevant to note that 7-space as origin fold, along its 5-space as a dimensional order, shall be leading from 7-space to 6-space as domain fold with 7-space as origin fold of the manifestation layer (4, 5, 6, 7).
19. With it at the next that is eight step, as the last step reach for 6-space domain shall be to its transcendental boundary (5-space in the role of boundary of 6-space) consisting of 12 components and hence the eighth and the final step reach to 5-space in the role of boundary with artifice value L = 12.
20. One may have a pause here and permit the transcending mind to remain in prolonged sitting of trans to be face to face with the phenomenon of spatial order 4-space in the role of dimension leading to solid order transcendental (5-space) domain in the role of boundary to comprehend and imbibe the features, order, values and virtues of the formulation 'INTERVAL' within creator space (4-space).

◆ ◆ ◆

15

GEOMETRY NVF (108)

1. The formulation GEOMETRY is of NVF value 108.
2. It as such is equivalent to NVF (END CREATION) = 108.
3. The formulation GEOMETRY, sequentially is of values (*i*). g = 7, (*ii*). E=5, (*iii*). O = 15, (*iv*). M = 13, (*v*). E = 5, (*vi*). T = 20, (*vii*). R = 18 and (*viii*). Y = 25.
4. These sequential values shall be leading us to
   (*i*) Starting with unity state/7-space/artifice 7 set-up
   (*ii*) As the 7-space is of transcendental (5-space) order in the role of dimension, as such it sequentially leads 7-space as domain to 5-space as dimension.
   (*iii*) (*iii*) The 5-space itself being of a solid order, as such, it is of linear equivalence as to be of artifice value $3 \times 5 = 15$ = NVF (O), the third phase and stage of sequential progression of the formulation GEOMETRY. (*iv*). As 5-space as solid dimensional order, it constitutes origin fold of spatial order creator space (4-space); and within creator space

(4-space) solids have a degree of freedom of motion, and as such 12 edged cube transits and transforms into 13 edged hyper cube, which as such becomes the fourth sequential phase and stage of the formulation (geometry). (*v*) Within creator space (4-space), the 13th edge gets connected with the transcendental (5-space) origin. As such the fifth sequential progression becomes of value '5' = NVF (E).

(*iv*) The transcendental flow through the origin of creator space (4-space) along artifices of sole syllable *Om* (ॐ) is a *Divya Ganga* flow of four fold values (9, 7, 3, 1) which from its both ends leading from 9-space to 7-space order and from 1-space to 3-space order, make it to be a flow of artifice values 20 = NVF (T).

(*v*) This flow manifests at the middle as of artifice value 18 = NVF (R). Here, itself it would be relevant to note that this manifestation as of four folds manifestation layer of solid order is of the expression (3, 4, 5, 6)/ (3-space as dimension, 4-space as boundary, 5-space as domain and 6-space as self referral origin)/(hyper cube 5 within creator space (4-space)

(*vi*) Hyper cube 5 as representative regular body of 5-space within spatial order creator space (4-space), as such shall be having a transcendental flow along each of the pair of dimensions of the spatial order, as such these together shall be transcendentally fulfilling the area within the dimensional axes as to be of artifices values $5 \times 5 = 25$ = NVF (AREA) = NVF (Y).

5. It would be blissful exercise to chase every individual formulations of NVF (25).
6. Here are being tabulated the bible formulations of NVF 25:

Flourish	Expounded
Executing	Falsifying
Excellency	Contendeth
Poorest	Pestilence
Perverse	Preserve
Petition	Ithrites
Counsels	Jealousy
Cupbearers	Impediment
Cuttest	Influences
Collection	Journey
Confirming	Jesuites
Devourer	Jesurun
Delivereth	Hamulites
Departure	Generation
Cyrenians	Gennesaret
Business	Gatherings
Cherethims	Fulfilling
Benefactors	Friendship
Approaching	Greatness
Asshurim	Worship
Artificers	Winebibbers
Afternoon	Youths
Accomplished	Wearisome
Loathsome	Vowedst
Maachathites	Uncomely
Lighteneth	Tychicus
Manslayer	Unblameably
Musicians	Troughs

Merarites	Twilight
Outlived	Utmost
Oppress	Unperfect
Oznites	Threshing
Paintedst	Tachmonite
Perisheth	Thickness
Perceiving	Therefrom
Pastors	Siphmoth
Plowers	Shulamite
Snuffers	Revilers
Shortened	Quarries
Suffereth	Samothracia
Surety	Scornful
Swearers	Scurvy
Sprinkled	Securely
Sorrow	Settest
Renderest	Servest
Repenting	Seraphims
Reproaches	

7. It would be blissful to chase sequentially the internal and external organizations for formulation of NVF (108).
8. The chase of the internal organizations of formulations of artifice 108, may be on the lines and of steps 1 + 107, 2 + 106, 3 + 105, and so on 54 + 54, in their both orientations as of the order 1 + 107, 107 + 1; 2 + 106, 106 + 2; and so on.
9. Here, It would be relevant to note that 108 = 54 + 54 leads to NVF (GEOMETRY) = NVF (SUN) + NVF (SUN).
10. Like that, it would be blissful to chase the formulations as follows:

(*i*) 108 = 54 + 54 = NVF (GEOMETRY)
= NVF (SUN) + NVF (SUN)

(*ii*) 108 = 55 + 53 = NVF (SKY) + NVF (AXIS)

(*iii*) 108 = 56 + 52 = NVF(LIGHT) + NVF (EARTH)

(*iv*) 108 = 57 + 51 = NVF (HEIGHT) + NVF (FULL)

(*v*) 108 = 58 + 50 = NVF (TWO) + NVF (VOID)

(*vi*) 108 = 59 + 49 = NVF (SOLID) + NVF (AXES)

One may have a pause and have a note as that the geometry as a discipline approaches the solar universe in terms of solid axes and reaches at sky – axis.

Continuing further,

(*vii*) 108 = 60 + 48
=NVF (GEOMETRY)=NVF (FOUR)+NVF(TREE)

It would be relevant to note that 4-space is a spatial order space within which transcendence is permissible at its origin and it as such leads to linear equivalence for 6-space as to be of value $6 \times 4 \times 2 = 48$.

Proceeding further,

(*viii*) 108 = 61 + 47
NVF (GEOMETRY)
= NVF (CHRUCH) + NVF (MIDDLE)

(*ix*) 108 = 62 + 46
= NVF (COSMIC) + NVF (LOGIC)

(*x*) 108 = 63 + 45
= NVF (LIMIT) + NVF (RANGE)

(*xi*) 108 = 64 + 44
= NVF (ZERO) + NVF (SPACE)

One may have a pause here and to fully comprehend and completely imbibe the values of the discipline of GEOMETRY by permitting the transcending mind to think, meditate and to transcend to be face to face with 0-space to glimpse its values and virtues.

Proceeding ahead

(*xii*) 108 = 65 + 43

NVF (GEOMETRY)

= NVF (CENTRE) + NVF (FRAME).

It would be blissful to glimpse the values and features of FRAME of the CENTRE.

(*xiii*) 108 = 66 + 42

NVF (GEOMETRY)

= NVF (FAMILY) + NVF (NEW)

The discipline of geometry would lead to new families of geometric set-ups.

(*xiv*) 108 = 67 + 41

= NVF (GEOMETRY)

= NVF (WATER) + NVF (AFFINE)

'Glimpsing' affine waters would be bliss of the geometricians.

(*xv*) 108 = 68 + 40

NVF (GEOMETRY)

= NVF (JOINT) + NVF (LINE).

By glimpsing the order and feature of 'joints of lines' one shall be enriching one's insight.

(*xvi*) 108 = 69 + 39

NVF (GEOMETRY)

= NVF (ZOOM) + NVF (ANGLE)

Geometrician glimpse the Angle of Zoom.

(*xvii*) 108 = 70 + 38

= NVF (STOP) + NVF (FIRE)

Geometrician reach where the fire stops.

(*xviii*) 108 = 71 + 37

= NVF (SPHERE)+ NVF (SEAL)

It would be blissful to be face to face with the seal of sphere, and also the way sphere seals the domains.

(*xix*) 108 = 72 + 36

= NVF (ORIGIN) + NVF (REAL)

It would be blissful to be face to face with real origin.

As such the chase of GEOMETRY infact is the chase of REAL ORIGIN.

(*xx*) 108 = 73 + 35

= NVF (FORMAT) + NVF (EYE)

(*xxi*) 108 = 74 + 34

= NVF (PAIRING) + NVF (ONE)

(*xxii*) 108 = 75 + 33

= NVF (SERIES) + NVF (SEED)

(*xxiii*) 108 = 76 + 32

= NVF (ADDITION) + NVF (LIFE)

It would be relevant to note that NVF (ADDITION) = NVF (MINUS) = 76

(*xxiv*) 108 = 77 + 31

= NVF (MATTER) + NVF (CAVE)

Geometrician chase the matter of within cave for enlightenment about the existence phenomenon along geometric formats.

(*xxv*) 108 = 78 + 30

= NVF (AMBROSIA) + NVF (BIBLE)

It is the ambrosia of the bible which feeds the intensity of urge to know the geometry.

It is a poser for the geometricians who answer whether 'life' can be added or subtracted.

(*xxvi*) 108 = 79 + 29
= NVF (NATURE) + NVF (BLACK)
Black nature is the heart of the discipline of geometry.

(*xxvii*) 108 = 80 + 28
= NVF (CREATOR) + NVF (AIR)
Creator of Air is ambrosial chase.

(*xxviii*) 108 = 81 + 27
= NVF (SQUARE) + NVF (HALF)
It would be blissful to work and chase the discipline of geometry as half 'square'. To half the square may look like obvious by having two parts of the square by drawing a diagonal within it. However, it would elude comprehension as to what would happen with the inclusion or exclusion of the diagonal itself while square would be attempted to be made of two parts. It is this feature which is a subject matter of focus here.

(*xxiv*) 108 = 82 + 26
= NVF (FACTORS) + NVF (GOD)
It would be blissful to make oneself conscious as to how the whole is put into parts which themselves are individually 'whole'. It is here where the formulation 'God' would come into play, for making FACTORS = A SQUARE. The formulation 'God' is of three sequential steps, namely, as of values '1, 7' (2, 15) and '3, 4'. The chase of these values would confront us as to how working with 'one' we reach 7 sequentially increasing hyper circles 1 to 7, and then, suddenly, at in between phase, at the middle of artifices 7 and 8, there is sudden inverse process coming into play and hyper circles 8 onwards starting decreasing. It as such would help us comprehend and to acquire insight as that infact whenever we try to split any manifested creations, these always split into three parts, though we try and desire to have it into two parts only it is this feature which takes us straight from 'one' to 'three' we reach straight a way

to a solid order it as such takes us to transcendental (5-space) domain of solid order and because of which we reach the next value after 'seven', as '15'=1 3 5. It is this jump from 7 to 15 within creator space (4-space), which is there being because of 'God', the creator space (4-space) value '4' = NVF (D).

Here, it also would be relevant to note that the *Upnishad* enlightenment for jump from 3-space to 4-space is of the sequential chase quadruple values (1, 2, 3, 8) taking into account the transition from linear order to spatial order which makes 4-space as to be of $4 \times 2 = 8$, linear sequential values equivalence for 4-space.

Further also It would be relevant to note that if we want to reach at the middle, first we reach at the first end, then at the second end and only there after at the middle of the two by constructing the middle point for the pair of given points. It would help us comprehend and to have an insight as to why letter 'D' for the formulation 'God' is at third place while other wise with $7 = 2\times3+1$ as geometries of 3-space and $15 = 1\times3\times5$ as dimensional order of 5-space would have brought 4-space in between 3-space and 5-space with sequential values order as 3-space, 4-space and 5-space. However the order as 7 for 3-space, 15 for 5-space has pushed 4 for 4-space at the end of the formulation.

(*xxiv*) 108 = 83 + 25

= NVF (FOLLOW) + NVF (AREA)

Geometry as such FOLLOWS AREA for the chase of the spatial order of creator space (4-space)

(*xxiv*) 108 = 84 + 24

= NVF (COLOUR) + NVF (EAR)

It would be blissful to note that the frequencies of colours make hearable sounds; NVF (HEAR) = NVF (LIFE)

NVF (HEAR) = NVF (H) + NVF (EAR)

= 8 EAR (S); 4 pairs of ears = all the years of four head lord/Lord Brahma/creator the supreme/overlord of 4-space/the overlord of the measure of the measuring rod of 6-space, which is of a hyper dimensional order. It would be relevant to note that 24= 6×4, equal to the dimensional order of 6-space.

(*xxiv*) 108 = 85 + 23

= NVF (END CREATION)

Here, it that way reaches the phase and stage of 'end creation' of the Discipline of Geometry.

(*xxiv*) 108 = 86 + 22

= NVF (PARAGRAPH) + NVF (GO)

To know geometry one shall 'go' to 'paragraph'.

(*xxiv*) 108 = 87 + 21

= NVF (GEOMETRY)

= NVF (TRUTH BEAM)

(*xxiv*) 108 = 88 + 20

NVF (GEOMETRY)

= NVF (CAGED VOLUME)

(*xxiv*) 108 = 89 + 19

NVF (GEOMETRY)

= NVF (EGG UNITY)

(*xxiv*) 108 = 90 + 19

NVF (GEOMETRY)

= NVF (ARTIFICES HEAD)

(*xxiv*) 108 = 91 + 17

NVF (GEOMETRY)

= NVF (MIRROR BACK)

(*xxiv*) 108 = 92 + 16

NVF (GEOMETRY)
NVF (REVERSE CAGE)

(*xxiv*) 108 = 93 + 15
NVF (GEOMETRY)
NVF (ENTITY FACE)

(*xxiv*) 108 = 94 + 14
NVF (GEOMETRY)
NVF (DEAD TRIMONAD)

(*xxiv*) 108 = 95 + 13
NVF (GEOMETRY)
NVF (RENEWING AGE)

(*xxiv*) 108 = 96 + 12
NVF (GEOMETRY)
NVF (BEE KNOWLEDGE)

(*xxiv*) 108 = 97 + 11
NVF (GEOMETRY)
NVF (PILGRIMAGE BED)

(*xxiv*) 108 = 98 + 10
NVF (GEOMETRY)
NVF (TRANSCEND BAG)

(*xxiv*) 108 = 99 + 09
NVF (GEOMETRY)
NVF (ADD THOUGHT)

(*xxiv*) 108 = 100 + 08
NVF (GEOMETRY)
NVF (CE DISCIPLINE)

(*xxiv*) 108 = 101 + 07
NVF (GEOMETRY)
NVF (BE ULTIMATE)

(*xxiv*) 108 = 102 + 06

NVF (GEOMETRY)
NVF (ABBA TWO SPACE)

(*xxiv*) 108 = 103 + 05
NVF (GEOMETRY)
NVF (COUNTING E)

(*xxiv*) 108 = 104 + 04
NVF (GEOMETRY)
NVF (FOUR SPACE D)

(*xxiv*) 108 = 105 + 03
NVF (GEOMETRY)
NVF (C DEFINITION)

It would be relevant to note that the third step of the sequence deserves to be defined as other wise like '1, 2' may take to '3' or to '4'.

(*xxiv*) 108 = 106 + 02
= (53 + 53) + (01 + 01)
= NVF (GEOMETRY)
= NVF (AXIS A, AXIS A)
= NVF (B INFINITY)

It would be relevant to note that 'A infinity' and 'B infinity', would be like 'infinity of line'/1-space/artifice 1 vis-à-vis infinity of surface/2-space/artifice 2.

(*xxiv*) 108 = 107 + 01
NVF (GEOMETRY)
= NVF (A REFLECTION)

It would be blissful to focus upon the 'reflection operation, which shall be bringing into focus the role of 'mirror' for the chase of the values of Discipline of geometry.

Further It would be relevant to note that ancient wisdom focus for exploration of space remains upon 'oils' and 'mirrors'; As such would be blissful to be face to face with;

NVF (OIL) + NVF (MIRROR) = NVF (GEOMETRY EGG). And as these number values formulations are availing 26 elements with element itself being NVF (ELEMENT) = NVF (PAIRING) And further as NVF (TWENTY SIX) = NVF (ULTIMATE TWO), the geometry deeply avails NVF (JOINT) = 68 = 34 + 34 = NVF (ONE) + NVF (ONE).

And with it geometry becomes 'pairing one' as the 'artifices and formats heads' run parallel to each other, as is the ancient wisdom enlightenment.

There are two established processing systems, namely *Sankhiya Nishtha* and *Yoga Nishtha* while the *Sankhiya Nishtha* presumes the existence of geometric formats and avails artifices of numbers and on the other *Yoga Nishtha* presumes existence of artifices of numbers and works out geometric format of dimensional spaces.

◆ ◆ ◆

16

VEDIC MATHEMATICS ANSWER

1. Poser

1. Vedic mathematics answer to the poser as to 'whether there can be a source format for whole range of alphabets of spoken languages?' is in affirmative.
2. However to chase it, first of all, it may be of interest to be through a practical illustration, in respect of our popular English Language itself.
3. In this respect as well, one of its world – formulation, namely 'IGNORANCE', on its chase would help deeper interest in the poser and the answer thereto.

2. Twenty Six Elements

1. Vedic systems prominently avail '26 elements'. For the present, these as such may be taken as artifices of range of numbers, '1 to 26'.
2. Further, at this stage, it also be taken as axioms as that '26 letter alphabet of English namely the letters range A to Z, being parallel to the numbers range '1 to 26'.

3. This, that way, may be taken as the range of numbers values formats for letters A to Z being 1 to 26, in that sequence and order. In short it may be taken as that NVF (A) = 1, NVF (B) = 2 and so on NVF (Z) = 26.
4. The formulation IGNORANCE conveys us the meaning.
5. It is not symbolic to have this formulation for its meaning as the formulation 'IGNORANCE' conveys.
6. Even the letters being availed and the sequence of availing these letters in the formulation 'IGNORANCE' are not a whimsical/ random choice from the alphabet letters range.
7. Our ignorance about this formulation in particular and English words (formulations) in general, shall be compelling us to have a pause and to reflect as to what is that sequence of values and format beneath which makes this formulation as of the meaning as this formulation conveys as being the 'IGNORANCE'.
8. The numbers values format chase for the letters availed by the formulation 'IGNORANCE', shall be at the outsets leading us to the sequential set-up of numbers '9, 7, 14, 15, 18, 1, 14, 3, 5'.
9. This numbers sequence '9, 7, 14, 15, 18, 1, 14, 3, 5' as such may not lead to meanings – sense but this number sequence has a definite meanings sense.
10. For it, for the present we may have another axiom as that 'elements' permit pairing of values which results into their summation because of which number value format for WORD would be equivalent to the sum of the number value formats of individual letters availed for the composition of the WORD.
11. This in case of the 'word' namely 'word' shall be leading us to, as that

    NVF (WORD) = NVF (W) + NVF (O) + NVF (R) + NVF

(D) = 23 + 15 + 18 + 4 = 60

12. It would be relevant to note itself that NVF (NUMBER) = 73 = NVF (FORMAT).
13. This as such shall be helping us leading to
    NVF (IGNORANCE) = 86 = NVF (PARAGRAPH)
14. Naturally, the poser would be as to how 'IGNORANCE' goes to be like 'PARAGRAPH'.
15. As '86' is a step ahead of '85', and as NVF (CREATION) = 85, so ignorance would be ignorance as it would be a step ahead of 'creation'.
16. This ignorance, is not only the end product but it also going to be of as many as of 9 sequential steps as to how there is to be a transition firstly from I to G, secondly from G to N, thirdly from N to O, fourthly from O to R, fifthly from R to A, sixthly from A to N, seventhly from N to C, eighthly from C to E and finally that is ninthly from E back to I.

3. About First Transition Step

17. The transition step from I to G is parallel to the transition from artifice 9 to artifice 7, which is of the format of transition from 9-space to 7-space.
18. Ancient wisdom enlightens us about artifice 9 as *Nav Braham* नव ब्रह्म.
19. *Nav* (नव) means nine, as well as *Nav* (नव) means 'new'.
20. Vedic mathematics would further help us transit from 9-space to 7-space as 9-space as domain fold to 7-space as dimension fold.
21. Therefore the basic ignorance, as such is to be as to how *Nav Braham* as domain fold is to be of unity state (7-space) dimensional order.
22. This ignorance, as such also shall be to remain as to how

the *Divya Ganga* flow comes to be of seven streams.

23. It also would manifest itself as an ignorance coming in the way of coordination of artifices '9 and 7' which respectively would be parallel to 9 geometries range of 4-space/9 versions of hyper cube 4. And 7 geometries range, 3-space/7 versions of cube.

24. Ahead it would to be the obstruction as to why with 9 as the biggest numeral of 10 place value system while 7 is the biggest prime numeral of this range.

4. About Second Transition Step

25. The ignorance at the second step would be of phase and stage of transition from letter G to letter N being availed in the formation 'IGNORANCE'.

26. This as such, would be of phase and stage of transition from artifice 7 to artifice 14.

27. Ancient wisdom enlightens us as that hyper cube 7 is the representative regular body of 7-space.

28. Further as that domain boundary ratio formulation for hyper cube 7 is A^7: 14 B^6.

29. It is this ignorance of domain boundary formulation for hyper cube 7 which is the manifested phase and stage of ignorance.

30. It would be relevant to note as that ancient wisdom has processed existence format pole star as of 7-space set-up while that of Sun being of 6-space set-up.

31. With 5-space playing the role of dimension of 7-space and the five head transcendental lord, Lord Shiv being the presiding deity of transcendental (5-space) domain set-up and their being precisely 14 *Maheshwara Sutras*, so it would help have insight about the degree of enlightenment needed to dispel ignorance of this phase and stage of the existence phenomenon.

5. About Third Transition Step

32. The ignorance at the third step would be of phase and stage of transition from letter N to letter O being availed in the formation 'IGNORANCE'.

33. This as such, would be of phase and stage of transition from artifice 14 to artifice 15.

34. It would be relevant to note the formulation 'NO', as it stands, as well in its meaning sense conveys as no.

35. NVF (NO) = NVF (BLACK). One may have a pause here and permit the transcending mind to think, meditate and transcend through this number value format of formulation 'NO' as of features 'BLACK'.

36. It is 'black' which shall not be obstructing the transition for the *Divya Ganga* flow of seven streams.

37. As such it is this obstruction which is to be at the base of the ignorance of this phase and stage.

38. It further would be relevant to note as that the start with phase and stage of the ignorance, which was of the feature of the transition from artifice 9 to 7/ 9-space to 7-space was infact for attaining transcendence through the in between '8-space'.

39. Further It also would be relevant to note that the flow which had emanated from *Nav Braham* source reservoir (9-space)/ artifice 9, and which had transcended through Asht Prakrati (8-space)/artifice 8, infact had been a systmal processing line of base/index reversal process, as much as that $9 = 3^2$ and $8=2^3$.

40. Therefore one shall have a pause again and to permit the transcending mind to remain in prolonged sittings of trans to be face to face with the features of obstructions in the flow process of this phase and stage, that is of transition from letter N to letter O/artifice 14 to artifice 15.

41. The artifice 14 also permits re-organization of four fold manifestation format (2, 3, 4, 5) / (2-space as dimension), (3-space as boundary) (4-space as domain), (5-space as origin).
42. The artifice 15 permits re-organization of features 15 $= 1 + 2 + 3 + 4 + 5 = 3 + 3 + 3 + 3 + 3 = 1 \times 3 \times 5$, as solid order domain.
43. As such the transition of this phase and stage as of the features of transition from artifice 14 to artifice 15 is of the features of transition from 5-space in the role of origin fold of 4-space to the role of 5-space as a domain fold.
44. It is this obstruction at which would be at the base of ignorance of this phase and stage.
45. One is to go to the shelter of the ancient wisdom to dispel this phase and stage of the ignorance as to how 5-space in the role of origin of 4-space shall be transiting and transforming into the role of 5-space as a domain fold.
46. It infact is the phenomenon of flow of transcendental values (of solid order) 5-space from origin fold of creator space (4-space) into the domain of creator space (4-space) itself.
47. One is to remain in prolonged deep sittings of trans to dispel ignorance of this phase and stage as to how the manifested values domain of creator space (4-space) gets fulfilled with the transcendental values of solid order transcendental (5-space) domain.

6. About Fourth Transition Step

48. The ignorance at the fourth step would be of phase and stage of transition from letter O to letter R being availed in the formation 'IGNORANCE'.
49. This as such, would be of phase and stage of transition from artifice 15 to artifice 18.

50. The formulation 'OR' has a meaning sense.
51. It is this meaning sense of the formulation 'OR', as 'or', which is, as to be of a choice, of this or that.
52. It is this choice making option which is the feature of 'intelligence' and as such to be that intelligent, the said phase and stage of ignorance is to be crossed over.
53. Ancient wisdom enlightens us as that the Prakrati/ (Nature) is of two values and virtues.
54. It is two fold values and virtués features make the *Prakrati*/ Nature as *Jad Prakrati* जड़ प्रकति/*mundane* nature and *Chetan* Prakrati/conscious Nature.
55. The flow which at the original start with phase and stage had jumped over Asht Prakrati, by this phase and stage is at the two fold choice junction, of which Jad Prakrati/mundane nature is their to obstruct and hence the ignorance of this phase and stage.
56. The living phenomenon is full of these two fold choice junctions, which may be designated in the context as of 'ignorance and knowledge gates/paths.
57. The tree with its feeder routes when are designated as *Jad* () mundane processing systems of single option. These in its operational functioning are though important feeder lines but are not the only feeder lines as the other feeder lines like that of Sun rays are there.
58. It would be relevant to note that NVF (SUN LIGHT)= 110 = NVF (MEDITATION) .
59. It is this conscious option choice which is to be availed to dispel with the ignorance of this phase and stage of the formulation IGNORANCE.
60. It would further be relevant to note as that NVF (OR) = 33 = NVF (SEED).
61. Seed is mundane, as well as life conscious reservoir.
62. Further it also would be relevant to note as that artifice

18, which is NVF (R), accepts re-organization as four fold manifestation as (3, 4, 5, 6)/ 3-space as dimension, 4-space as boundary, 5-space as domain and 6-space as origin.

63. One may have a pause here and note as that this four fold manifestation (3, 4, 5, 6) is the set-up of hyper cube 5.

64. Ancient wisdom enlightens as that hyper cube 5 being the representative regular body of 5-space manifests within creator space (4-space) as a four fold manifestation layer (3, 4, 5, 6) with 6-space as its origin fold.

65. Further as that 6-space in the role of origin fold, itself is of hyper dimensional order i.e. 4-space being in the role of dimension.

66. It is because of this hyper dimensional order for the origin fold which with transcendental flow within (it) that is within the hyper dimensional order transforms the letter into syllables.

67. It would be relevant to note as that NVF (LETTER) = 80 = NVF (CREATOR). And NVF (SYLLABLE) = 88 = NVF (VOLUME) = NVF (SPACE, SPACE).

68. As such it is this feature of transition and transformation for letters into syllables which would help dispel ignorance of this phase and stage.

7. About Fifth Transition Step

69. The ignorance at the fifth step would be of phase and stage of transition from letter R to letter A being availed in the formation 'IGNORANCE'.

70. This as such, would be of phase and stage of transition from artifice 18th letter to the first letter, that is back to the first letter transformed into first syllable.

71. Lord Krishna, incarnation of Lord Vishnu, the overlord of 6-space, in *Shrimad Bhagwad Geeta*, enlightens as that

'amongst syllables I am *Akara*'.

72. This enlightenment would dispel ignorance about the transition and transformation of the letters into syllables and consequential cyclic orders for the systems within transcendental (5-space) domains making it to be of the range of artifices values 1 to 5/ measuring rod of hyper cubes 1 to 5 and a step ahead the cyclic order come into play and the system to start processing again with sixth step being the first step.

73. *Shrimad Bhagwad Geeta* organization format, as such accepts same range of *sholokas* for chapter-1 as well as for chapter-6.

8. About Sixth Transition Step

74. The ignorance at the fifth step would be of phase

75. and stage of transition from letter A to letter N being availed in the formation 'IGNORANCE'.

76. This as such, would be of phase and stage of transition from artifice 1 to artifice 14.

77. The ignorance this phase and stage as such would be as to why the range of this phase ignorance is of the range of artifices 1 to 14.

78. Here, It would be relevant to note that it is because of the transition and transformation for the letters into syllables, as that the seven streams *Divya Ganga* Flow of the start with position, here would get multiplied as of value '7× 2=14'.

79. Further it would relevant to note as that the syallable '*Akara*' being of the values and virtues of the (6-space), it shall be making this range 'An', as to be of the range of the boundary of 7-space/hyper cube 7.

80. Still further, it also would be relevant to note as that the artifice 14 permit re-organization as four fold

manifestation layer (2, 3, 4, 5).

81. Still further It would be relevant to note that that the transition of this and the previous phase and stage together shall be of the features of the formulation RAN, which shall be conveying the meanings as it conveys as ran.

82. Also this formulation 'RAN' shall as well be leading to the formulation 'AN', with the meanings as it conveys as 'an'.

83. The transition from the phase and stage of letter 'R', to letter 'N', as such would be the transition and transformation from the features of artifice 18 to that of artifice 14.

84. It would be relevant to note that the transition and transformation from artifices 18 to 14 as such shall be of the order and features of four fold manifestation layers (3, 4, 5, 6) and (2, 3, 4, 5, 6).

85. One may have a pause here and permit the transcending mind to remain in prolonged sittings of trans and to glimpse the transcendence phenomenon happening within the dimensional orders, which within the transcendental (5-space) domain shall be leading to 4-space as origin fold of 3-space in the role of dimension.

86. Here, it would be relevant to note that 4-space being the origin of 3-space, would make it a spatial order origin for a solid dimensional order of transcendental (5-space) domain. And, this as such shall be helping have an insight into the meanings of the formulation 'RAN'.

9. About Seventh Transition Step

87. The ignorance at the fifth step would be of phase and stage of transition from letter N to letter C being availed in the formation 'IGNORANCE'.

88. This as such, would be of phase and stage of transition from artifice 14 to artifice 3.

89. The organization for the artifice 14 as spatial order manifestation layer (2, 3, 4, 5) shall at a step ahead would leading to the solid order manifestation layer, and comprehension of this transition, as such shall be helping to dispel the ignorance of this phase and stage of the formulation 'IGNORANCE'.

10. About Eighth Transition Step

90. The ignorance at the fifth step would be of phase and stage of transition from letter C to letter E being availed in the formation 'IGNORANCE'.

91. This as such, would be of phase and stage of transition from artifice 3 to artifice 5.

92. 3-space playing the role of dimensional order for 5-space, when chased as such shall be helping dispel the ignorance of this phase and stage of the formulation 'IGNORANCE'.

11. About Ninth Transition Step

93. The ignorance at the fifth step would be of phase and stage of transition from letter E to letter I being availed in the formation 'IGNORANCE'.

94. This as such, would be of phase and stage of transition from artifice 5 to artifice 9, that is back to start with position of the formulation itself.

95. One may have a pause here and permit the transcending mind to remain in prolonged sittings of trans to chase and rechase the above phases and stages of ignorance of the formulation 'IGNORANCE' and to comprehend and to have an insight as to how the transcendental order runs through this whole range.

96. It would be relevant to note that artifice 5 is at the middle of the artifices range '1 to 9'.

97. Still further It would be relevant to note that NVF (PARAGRAPHED) = 86 + 9 = 95 = NVF (RENEWING).

98. It is this renewing feature which shall be taking us back from NVF (E) = 5 phase and stage to the phase and stage of NVF (I) = 9.

99. It would further be relevant to note that

NVF (FIVE) = 42 = NVF (NINE) = NVF (NEW).

100. It infact is a transition attainment and as such a breakthrough from NVF (OLD) = 31 = NVF (CUBE), a representative regular body of 3-space, a linear order space to NVF (NEW) = 42, a spatial order set-up.

101. The fresh start from the first letter 'I' of the formulation 'IGNORANCE', as a spatial order start, as that $9 = 3^2$ till the end up till letter E = 5/5-SPACE shall be attaining a solid order, and then the cyclic restart, naturally shall be taking from solid order to hyper solid order, and the process shall go on renewing itself infinitely repelling ignorance at its each and every phase and stage.

◆ ◆ ◆

SECTION – 2: SPECIFIC FEATURES

17. Transcendental Boundary
18. Dwadash Adityas दवादस अद्वितीय्
19. 5-space within 4-space
20. 6-space within 5-space
21. 7-space within 6-space
22. Pole Star as Origin of Solar Universe
23. 7-space as Origin of 6-space
24. First Vowel
25. Second Vowel
26. Third Vowel
27. Fourth Vowel
28. Fifth Vowel
29. Sixth Vowel
30. Seventh Vowel
31. Eighth Vowel
32. Ninth Vowel

17
TRANSCENDENTAL BOUNDARY

1. The transcendental boundary ((5-space ⑤) of self-referral space (6-space ▷◁) is of 12 components parallel to 12 months of a year.
2. The values and order of 12 components of transcendental boundary are parallel to values and order of 12 names of Lord Ganesha ●.
3. Twelve transcendental names of Lord Ganesha ● are *Sumukha, Ek-danta, Kapila, Gajkarna, Lambodra, Vikate, Vighan nashna, Vinayaka, Dhurmketu, Ganadhiksha and Gajananna.*
4. The values and order of twelve transcendental component of transcendental boundary are parallel to values and order of 12 syllables of *Dharuv mantra* (ओम नमोः भगवते वासुदेवायः) *Om Namoh Bhagwate Vasudevaeye'*).
5. These values and order are the values of the *Dwadash Adityas* (12 Suns).
6. Within each of the transcendental component of the transcendental boundary, a transcendence takes

place which its upper half/northern upper half, and simultaneously it also permits ascendance from (1-space –) to (3-space □) to (5-space $) in its lower half/southern lower half.

7. Further, at the middle centre/origin source there being a self-referral seat (6-space ⊳⊲) of hyper dimensional/creative order (4-space ⌑) and as the origin of creative dimension there being a transcendental seat, as such there emerges a transcendental flow within each of the six dimensions of (6-space ⊳⊲).
8. One shall sit comfortably and permit the transcending mind to remain in prolonged deep trans to be face to face with the simultaneous play of *Trimurti* [(4-space B), (5-space $) and (6-space ⊳⊲)'] .
9. This as such brings into operation the values and order of 4-space mathematics, 5-space mathematics and 6-space mathematics.
10. It would be relevant to note that the values of sequential order (1, 1,——), (1, 2, 3,—), (1, 2, 4,—) and manifestation layer format (1, 2, 3, 4), transcendence range format (1, 2, 3, 4, 5) and self-referral base (1, 2, 3, 4, 5, 6).

◆ ◆ ◆

18

DWADASH ADITYAS दवादस अद्वितीय्

1. Sun, as (6-space) body with (4-space) in the role of dimension gives values and order for the dimensional frame as of value $6 \times 4 = 24$ which is parallel to 24 hours of the day.
2. (6-space) with (5-space) in the role of boundary is of the values $6 \times 5 = 30$ parallel to 30 days of the month.
3. Hyper cube 6 as representative regular body of (6-space) accepts 12 components boundary, which is parallel to the 12 months of a year.
4. This is the creative dimensional order (4-space in the role of dimension) mathematics which settles the age of Lord Brahma (creator the supreme) as 432000 years.
5. The age of Lord Brahma as $432000 = [\{(400 + 4000 + 400) + (300 + 3000 + 300) + (200 + 2000 + 200) + (100 + 1000 + 100)\} \times 360]$
6. With the emergence of transcendental values from within the creative dimensions, the same with their flow through the six dimensions of (6-space),

equip the transcen-dental boundary components with creative transcen-dental order, with which each of the transcendental components gets potentialized as the values and order of Sun and thereby the Sun multiplies as 12 folds i.e. 12 Suns/*Dwadash Adityas*.

7. It would be relevant to note that Lord Brahma, creator the supreme, presiding deity of (4-space) with meditation in the cavity of his heart upon Lord Shiv, the Lord of transcendental worlds (5-space) multiplies 10 folds as 10 *Brahamas* as the boundary of *Shiv Lok* (5-space).
8. As such the phase and stage ahead, is the phase and stage of transcendental meditation of Lord Shiv within cavity of his heart on Lord Vishnu and Lord Shiv multiplies 10 fold as 12 Shiv of the values and order of Lord Vishnu, 12 in number.
9. One shall sit comfortably and permit the transcending mind to be phase to phase with the meditation attainment of Lord Brahma multiplying as 10 *Brahamas*.
10. And further to be face to face with transcendental meditation of Lord Shiv multiplying as *Dwadash Adityas*.

◆ ◆ ◆

19

5-SPACE WITHIN 4-SPACE

1. One shall sit comfortably and permit the transcending mind to chase the phenomenon of (5-space) within (4-space).
2. This would be the phenomenon of (5-space) accepting (4-space) as creative boundary.
3. It further would be a phenomenon of origin of (4-space) being of transcendental seat of the values and order of (5 space).
4. One shall sit comfortably and permit the transcending mind to remain in prolonged deep trans to simultaneously chase the outward and inward transcendental progression within creator space (4-space).
5. It would be blissful exercise to chase (4-space) in its different role, as well as (5-space) in its different role and simultaneous existence phenomenon of (4-space) and (5-space) in manifestation layer, transcendence ranges and self-referral base.

◆ ◆ ◆

20

6-SPACE WITHIN 5-SPACE

1. One shall sit comfortably and permit the transcending mind to chase the phenomenon of (6-space ▷◁) within (5-space □).
2. This would be the phenomenon of (6-space ▷◁) accepting (5-space □) as creative boundary.
3. It further would be a phenomenon of origin of (5-space $) being of transcendental seat of the values and order of (6-space ▷◁).
4. One shall sit comfortably and permit the transcending mind to remain in prolonged deep trans to simultaneously chase the outward and inward transcendental progression within creator space (5-space $).
5. It would be blissful exercise to chase (5-space $) in its different role, as well as (6-space ▷◁) in its different role and simultaneous existence phenomenon of (5-space $) and (6-space ▷◁) in manifestation layer, transcen-dence ranges and self-referral range and unity basis.

◆ ◆ ◆

21

7-SPACE WITHIN 6-SPACE

1. One shall sit comfortably and permit the transcending mind to chase the phenomenon of (7-space) within (6-space).
2. This would be the phenomenon of (7-space) accepting (6-space) as self-referral boundary.
3. It further would be a phenomenon of origin of (6-space) being of unity state seat of the values and order of (7-space).
4. One shall sit comfortably and permit the transcending mind to remain in prolonged deep trans to simultaneously chase the outward and inward transcendental progression within self-referral progression within unity state (6-space).
5. It would be blissful exercise to chase (6-space) in its different role, as well as (7-space) in its different role and simultaneous existence phenomenon of (6-space) and (7-space) in manifestation layer, transcendence ranges and self-referral range and unity state with Asht Prakrati as base.

◆ ◆ ◆

21

POLE STAR AS ORIGIN OF SOLAR UNIVERSE

1. One way to chase (7-space) Pole star is as a unity state origin of (6-space)/Sun.
2. The other way to chase pole star is as a unity state measuring rod (1, 2, 3, 4, 5, 6, 7) / ((1-space –), 2-space, (3-space □), (4-space B), (5-space), (6-space) (7-space)'/hype cube 1 to 7.
3. Still other way to chase pole star (7-space) is along artifice 28, the second perfect number.
4. The artifice 28 accept re-organization as 28 = 4   7.
5. It would be relevant to note that (4-space) has transcendental seat (5-space) at its origin and that (5-space) plays the role of dimension of (7-space).
6. Artifice 28 also accepts re-organization as [(1 + 2 + 4) + 7+14].
7. It is a five steps long progression of values and order 1, 2, 4, 7 and 14.
8. This is of the values and order (20, 21, 22) + (20 , 21, 22).

9. It would be blissful to chase this five steps long progression along five dimensional frame of (7-space).
10. One shall sit comfortably and permit the transcending mind to remain in prolonged deep trans to be face to face with this unity state existence phenomenon.

◆ ◆ ◆

23

7-SPACE AS ORIGIN 6-SPACE

1. 7-space as origin of (6-space ⊳⊲) is the phenomenon of attainment of unity state within (6-space ⊳⊲)/Sun.
2. As (5-space ＄) plays the role of boundary of (6-space ⊳⊲) as well as dimension of (7-space ⩚), as such it would be a blissful exercise to chase this phenomenon of (5-space ＄) at the boundary attaining (7-space ⩚) within (6-space ⊳⊲) as well as in the universe sphere around (6-space ⊳⊲).
3. (7-space ⩚) as origin also deserves to be chase as a seven stream flow from the *Bindu Sarovar* Q.
4. It also deserves to be chased as of the format of seven geometries of (3-space □).
5. Further it also deserves to be chased as seven non negative geometries of (6-space ⊳⊲).
6. Still further it deserves to be chased as seven non positive geometries of (6-space ⊳⊲).

7. Still further it deserves to be chased as manifestation layer (4, 5, 6, 7).
8. Still further it deserves to be chased as transcendental range (3, 4, 5, 6, 7).
9. Still further it deserves to be chased as self-referral range (2, 3, 4, 5, 6, 7)
10. It would be a blissful exercise to chase the transition and transformation from (6-space ) to (7-space ) as a transition and transformation from 6 as first perfect number to 28 as second perfect number.

◆ ◆ ◆

24
FIRST VOWEL

1. *'Akara'* is the first vowel.
2. is the *devnagari* script form of *'Akara'*.
3. The features of this script form are two fold as of a cube placed upon four legged table.
4. This script form, as of these pair of features, makes the *Devnagri* alphabet as of a proper start to proceed from 3-space/ (cube) to creator space (4-space).
5. It is an expression of manifested creations; 'domain' '3-space/ ' with 'base ' ((4-space))
6. Manifested creations set-up (3-space/A) is a linear order set-up with spatial order base.
7. It is parallel to the set-up of a 3-space/ as domain and (4-space) as origin.
8. As such, it is parallel to the four fold linear order manifestation format as linear dimensional order, spatial boundary, solid domain and hyper solid origin.
9. Parallel to it would be the quadruple (1, 2, 3, 4).

10. One shall sit comfortably and permit the transcending mind to have a fresh look at the script form of first vowel of *Devnagri* alphabet/*'Akara'* .

'1'

1. *Akara* being the first vowel, as such it is to accept the values and order of '1'.
2. '1' as artifice 1, as 1-space, as an interval being the representative regular body of 1-space as linear order, as a manifestation layer and like, there are whole range of features and values which of their own get associated with the first vowel/*Akara*.
3. '1' is first after '0' and as such '1', may be both '0 and 1' and whole range in between.
4. First vowel, as such, with conscious choice application shall be for the whole range '0 to 1' as well as both are either of the end values '0' or '1'.
5. One shall sit comfortably and permit the transcending mind to be face to face with the last Sutra of *Ashtadhey* (grammar book) ''.
6. One way to approach it is '0' is '1'.
7. As such '1' is whole, as well as a part of the whole.
8. The letter ' as such merges with consonants.
9. Further '+' as of parallel vertical lines format deserves to be chased.
10. Still further 0 + 0 = 0 for '+', as synthesis of smaller lines as a bigger line as well deserves to be chased as a track of moving point.

1-Space

1. One shall sit comfortably and permit the transcending mind to chase line as a track of a moving point.
2. With this format, 1-space itself can be chased as a track of a 0-space.

3. Line/interval provide sequential placements format.
4. It is this feature of 1-space/line/interval/first vowel ", which shall be making first vowel being at the core of the values of all the vowels.
5. Infinite line, at infinity transiting and transforming as linear boundary of spatial set-up brings to focus the limit of expression of line/1-space/first vowel as a range of single digit numbers (numerals).
6. Nine numerals range for ten place value system, as such shall be settling full expression for the line as expression range of nine vowels/nine numerals.
7. It would be blissful exercise to permit the transcending mind to chase '9 as 1, 8 as 1, 7 as 1, 6 as 1, 5 as 1, 4 as 1, 3 as 1, 2 as 1 and 1 as 1'.
8. Further, It would also be a blissful exercise to chase '1 as 1, 1 as 2, 1 as 3, 1 as 4, 1 as 5, 1 as 6, 1 as 7, 1 as 8 and 1 as 9'.
9. Likewise, it would be a blissful exercise to have a transition from first vowel to second vowel to third vowel, third vowel to fourth vowel, fourth vowel to fifth vowel, fifth vowel to sixth vowel, sixth vowel to seventh vowel, seventh vowel to eighth vowel, eighth vowel to ninth vowel.
10. And other way around as well, it would be a blissful exercise to chase transition from ninth vowel to eight vowel, eighth vowel to seventh vowel, seventh vowel to sixth vowel, sixth vowel to fifth vowel, fifth vowel to fourth vowel, fourth vowel to third vowel, third vowel to IInd vowel, and IInd vowel to first vowel.

Interval

1. One shall sit comfortably and permit the transcending mind to chase the set-up of an interval.

2. One of the prominent feature of the interval is its orientation from one end point to another end point.
3. Along infinite line, it shall be having a pair of opposite orientations viz. from first end point to second end point and second end point to first end point, both constituted a reflection pair.
4. Within a plane/2-space, interval shall be having as well as eight orientations, along sides and diagonals of the square through its corner points and through the centre.
5. It would be relevant to note that 'curve' shall be of in between placement/feature/values of 1-space at one end and 2-space at another end.
6. With this in between state, a bridging for the gap in between the features of (1-space) set-up and (2-space) set-up can be attained.
7. The curve is a set-up of points, and as such is a track of a moving point, like state line being a track of a moving point but with the difference as that here the motion, liberty for the point is of the values of liberty of motion in a surface.
8. This way curve is a line but of a surface format.
9. It is this which brings in the *Anstha* letters.
10. It would be blissful exercise to permit the transcending mind to chase the script form of letter *Akara* as vertical line, as well as a curve being a reflection pair of the first *anthstha* letter of *Devnagri* format, namely '*yakara* य्'.

Linear Order

1. (1-space –) as a track of a moving point is like 0-space/points having placement upon 1-space/line.
2. And thereby 0-space constituting boundary of (1-space –).

3. (1-space –) as dimension of 3-space/ imbeds linear order in

   3-space/ as domain.

4. This, as such makes (1-space –) in the role of dimension of 3-space/ .
5. 0-space likewise plays a role of dimension for 2-space.
6. As such, simultaneous availability of 0-space and 1-space, in the format of a line, as a track of a moving point and also with the potentialities to play the role of dimension, shall be making sequencing for points along line in the role of dimension as formats for the sequential set of nine vowels of *Devnagri* alphabet.
7. It would be like a synthetic measuring rod of representative regular bodies of (1-space –), (2-space ■), 3-space/ , (4-space), 5-space/ , (6-space), (7-space)', 8-space and 9-space.
8. This as such, shall be providing sub ranges/sub measuring rods for the domain of (2-space ■) as set-up of interval and square, for 3-space/– the set-up of interval, square and cube and so on.
9. Parallel to it, would be the coordination of first two vowels, first three vowels, first four vowels and so on.
10. Of these, most blissful exercise would be to reach at the values of synthesis of first vowel with first vowel, of first vowel with second vowel, of first vowel with third vowel, of first vowel with fourth vowel and so on.

◆ ◆ ◆

25

SECOND VOWEL

1. Transition from first vowel to second vowel is parallel to transition from '1' to '2'.
2. It is also parallel to transition from '1-space' to '2-space'.
3. Likewise it would be parallel to transition from interval to square.
4. One shall sit comfortably and permit the transcending mind to have a fresh look at the script form of second vowel, namely, *Ekara* (इ).
5. Amongst others, one of the feature of this script form is as of a drilling screw.
6. It would be blissful exercise to permit the transcending mind to chase the transition from script form of first letter *akara* '' being cube cube/solid upon four legged table/with base to the script form of second vowel *Ekara* as a drilling screw focusing upon the next internal progression step for domains being to drill through the origin seal of the manifested domains.

7. Parallel to it would be a chase at the centre of the grid frame for the surface.
8. It would be like being of the values and order of A^2 as unit/unit area as comparison to $A^1 = 1$ = unit length format for the first vowel.
9. It would be a blissful exercise to simultaneously chase pair of first two vowels as interval and square simultaneously existing within a square.
10. It would be like at infinity '1' transiting as '2', and they together as '3' making the whole set-up as '6' parallel to sixth vowel, accepted as a grammar rule 'अ + ई = ऐ'

'2'

1. '2' as '1' and '1' as '2' when chased shall be giving inside of the structural values differences of both these artifices, and parallel to it one shall be having inside about the differences of features and values of first and second vowel.
2. Likewise the chase 1^1 to 1^2 shall be enriching about the structural differences of measures of length and area.
3. '1' as '0 and 1' and '2' as '1 and 2' as well shall be providing additional insight about the structural feature of interval and square/1-space and 2-space/ first interval and second interval.
4. Points as boundary of interval and interval as boundary of square, as well deserve to be chased to reach at the structural feature of interval and square as manifested bodies of (1-space –) and (2-space ■) respectively as formats for the features of first vowel and second vowel.
5. Along linear format chase for first vowel and second vowel may be in terms of quadruples (1, 2, 3, 4) and (2, 3, 4, 5) .

6. However along, spatial format the chase for the roles of first vowel and second vowel may be in terms of 4 4 matrix formats as under

A for first vowel

1	2	3	4	–2	–1	0	1
0	1	2	3	–1	0	1	2
–1	0	1	2	0	1	2	3
–2	–1	0	1	1	2	3	4

B for second vowel

2	3	4	5	–1	0	1	2
1	2	3	4	0	1	2	3
0	1	2	3	1	2	3	4
–1	0	1	2	2	3	4	5

7. It would be blissful exercise to chase first vowel in different roles parallel to artifice '1' and second vowel parallel to artifice '2'.
8. Further it would be blissful exercise to chase transition for the roles of first vowel into the roles of second vowel along 4×4 matrix format.
9. It would further be blissful exercise to chase first vowel and second vowel respectively along the respective diagonals of 4×4 matrix format.
10. Still further it would be very blissful exercise to chase first vowel and second vowel along the pair of orientations as placements as diagonals of 4×4 matrix format depicted above.

2-Space

1. (2-space ■) has pair of dimensions.
2. As 0-space plays the role of dimension of (2-space E), as such, it would be a blissful exercise to chase synthesis of pair of dimensions of (2-space ■) as $0 + 0 - (-2) = 2$.

3. It is this jump over (1-space –) which deserves to be chased thoroughly.
4. Likewise the jump over ((2-space ■) as well deserves to be chased thoroughly.
5. And, ahead (2-space ■) in the role of dimension will jump over 3-space /  to reach creator space (4-space).
6. It is this phenomenon of dimensional synthesis of domains which deserves to be thoroughly chased.
7. Parallel to it would be the phenomenon of jump over second vowel at the base of *Aum* (ओम्) formulation.
8. It would be relevant to note that (2-space ■) within 3-space / A would get framed within a two dimensional frame of pair of linear dimensions.
9. It would be blissful exercise to distinguish (2-space ■) of a pair of dimensions of 0-space value each from that of (2-space ■) within a pair of linear dimensions.
10. One shall sit comfortably and permit the transcending mind to distinctively chase (2-space ■) within a pair of dimensions of 0-space value each and pair of linear dimensions of (1-space ■) value each.

Square

1. Square is a representative regular body of (2-space ■).
2. Likewise circle is also a representative regular body of (2-space ■).
3. One common feature of the set-ups of square and circle is that their domain and boundary accept the formulation $A^2 : 4B^1$.
4. This is in continuity of the domain – boundary structural formulation for interval being $A^1 : 2B^0$.
5. It is this common formulation $A^N : 2NB^{N-1}$ for N = 1, 2, 3, 4, — which is distinguishing *Devnagri* vowels formats from other letters of *Devnagri* alphabet.

6. It is under above structural feature that (2-space ■) accepts grid frames for the grid zones.
7. It would be a blissful exercise to transit from upper face of the surface/plane/square/lower surface.
8. Still further it would be a blissful exercise to chase 3-space / A as a track of a moving (2-space ■),
9. Still further It would be very blissful exercise to approach 3-space/⧉ as a gap between the pair of parallel planes.
10. Still further it would be a blissful exercise to approach 3-space/⧉ within spatial envelop stitched by six planes.

Spatial Order

1. Spatial order is of very distinct features that that of linear order.
2. Linear order accepts length as a unit vide spatial order accepts area as a unit.
3. Cube is the representative regular body of 3-space/⧉ as a set-up of linear dimensional order.
4. The prominent feature of this set-up of linear dimensional order is that 3-space/⧉ accepts seven geometries and parallel to it there are seven versions of cube.
5. It is because of this feature that hyper circles increase only up till hyper circle-7 and hyper circle 8 onwards there would be increase of circumference value.
6. Spatial order over comes this break.
7. It is because of hyper cube 4 accepting solid boundary of eight components.
8. It is this feature of spatial order which works out sequentially increasing boundaries for all the represen-

tative regular bodies of dimensional spaces, which deserves to be chased thoroughly for full comprehension and imbibing of these features of spatial order.

9. It is this feature of spatial order which distinguishes second vowel from the first vowel and infact from all other vowels.
10. Further it is because of this feature that (4-space ) with spatial dimensional order becomes the creator space (4-space) for the whole range of manifested existence phenomenon.

◆ ◆ ◆

26
THIRD VOWEL

1. Third vowel of *Devnagri* alphabet is *Ukara* (उ).
2. This script form of *Devnagri* alphabet letter *Ukara* (उ) deserves to be chased in continuity of the sequential steps of script forms of letters *akara* and letter Ekara (इ).
3. Letter *akara* is a cube upon table (domain with base) and letter *Ekara* is a drill screw making it possible to reach the centre/origin of the domain of cube and even to have drill a step ahead through the origin.
4. This shall be resulting into a split of the domain into a pair of hemispheres, and these together, as such shall be synthesizing the script format for letter *Ukara* (उ).
5. These steps together complete a phase and stage of the chase, as such, a set of first three vowels, and these together stand coordinated as *Maheshwara Sutra*-1 (अ, इ, ए, ण्)
6. One shall have a fresh look at this coordination in terms of the script form of letter (ण्) designated as *Anubandh*.
7. One meaning of *Anubandh* is 'to get bound by following'.
8. By following the script format of (ण्), one may be compre-hending as a chase path of first step being

a vertical downward flow, at second step a linear horizontal flow and finally at third step, as a vertically upward flow.

9. This as such shall be fixing the sequential coordination for first three vowels as of linear vertical downward flow path for letter *akara*, spatial horizontal flow path for letter *Ekara* and solid vertically upward flow path for letter *Ukara*.
10. This as such, also settles the coordination for interval, square and cube, in a way to envelope the space as is the script format for letter (पु).

'3'

1. Artifice '3' follows artifices 1 and 2 and parallel to it follows third vowel, the first and second vowels.
2. Parallel to artifice '3' is 3-space, and in a sequence, likewise third vowel avails the format of 3-space.
3. Cube as representative regular body of 3-space/A, as such, it works out the flow path for third vowel as coordinated by first *Maheshwara Sutra*.
4. Like cube, sphere is also the representative regular body of 3-space, and as such its split into a pair of hemispheres helps settle the script format for third vowel.
5. 3-space/ accepts three dimensional frame of linear dimensions parallel to three artifices of whole number 3.
6. These three dimensions, sequentially as a single dimension, pair of dimensions and all the three dimensions provide coordination format for first three vowels as a first *Maheshwara sutra*.
7. One shall sit comfortably and permit the transcending mind to chase interval, square and cube within a cube.
8. One shall have a fresh look at the above coordination feature of three vowels as first *Maheshwara Sutra* as interval, square and cube coordinated within cube following the script format of letter *Ukara*.

9. One way to look at the set of first three vowels is as linear dimension, spatial boundary and solid domain of cube.
10. The other way to look at the set of three vowels is as (1-space –) in the role of dimension, (2-space ■) in the role of boundary and 3-space/ in the role of domain manifesting together.

3-Space

1. 3-space being third following 1-space and 2-space becomes self sustained as a domain availing its all the three dimensions.
2. It is this feature of 3-space which makes him of prominent feature of a domain fold of the four fold manifestation layer within creator space (4-space).
3. 3-space as domain fold accepts its origin in a sealed state.
4. It is this sealed state of the origin which makes it (origin) as being in distinguishable from all other points of the domain (3-space).
5. Origin becoming as an ordinary point of the domain is a feature which deserves to be comprehended fully.
6. It is this feature as that the origin of the domain becomes the ordinary domain point which makes the domain as a self contained set-up.
7. It is because of this self-contained feature of the domain/here 3-space, which makes it as a self contained existence phenomenon.
8. This being so, there arises a need to workout a process to transcend through the domain, and for it the ancient wisdom avails the halving process of dimensional frames, and as such a three dimensional frame of half dimension, which provides script format for *Ushmana* letters (*Sakara* स्) of the prominent role and also *Ukara* gets its half as a *matra* measure as (ु).
9. This would help us appreciate the role of 'सु'.
10. Further it would also help us appreciate the role of 'सु'

with 'स्' removed, making 'सु' following first vowel, as to be of the features of ':' / *visergnia*.

Cube

1. For cube to transcend through 3-space as domain fold is to transcend through its centre/origin and for it, its seal is to melt to give way to creator space (4-space) sealed at origin seat.
2. With it, the cube shall be acquiring a degree of freedom of motion within 4-space.
3. With it cube, in a dynamic state, shall be at the boundary of 4-space.
4. It is this acquired feature of cube transiting from static to dynamic state, which deserves to be chased.
5. As, with this acquired feature, infact, on the cube though of a linear order, shall be transiting into a spatial order and with it shall be transforming its role from that of a static domain to a dynamic boundary state.
6. While chase this phenomenon of transition from static domain state of cube to dynamic boundary state of cube, one shall ever remain conscious as that infact the whole phenomenon has of its own silently transited from 3-space to 4-space, which of a spatial dimensional order.
7. One shall have a pause here and permit the transcending mind to remain in deep trans to fully comprehend and imbibe this transition phenomenon.
8. One shall have a fresh look at the set-up of the cube and comprehend its feature as that in its all the eight corner points are imbedded three dimensional frames of half dimensions, and that is why the centre of the cube as origin of 3-space permit split for the cube as eight sub cubes parallel to eight octants cut of 3-space and split for boundary of hyper cube 4 as of eight solid components.
9. Parallel to it, one shall attempt chase for the set-up of a sphere.

10. It would be blissful exercise to chase split of a sphere into a pair of hemisphere and these together re-synthesizing as a script format for third vowel *Ukara* (उ).

Solid Order

1. 3-space in the role of dimension structures a solid order for 5-space/𝔖.
2. One shall sit comfortably and permit the transcending mind to chase the set-up of cube availing five three dimensional frames.
3. Four of the three dimensional frames availed by the set-up of the cube split and provide eight three dimensional frames of half dimensions.
4. These eight three dimensional frames of half dimensions are imbedded into eight corner points of the cube.
5. One three dimensional frame (of full dimension) is imbedded at the centre of the cube.
6. One shall sit comfortably and permit the transcending mind to chase eight three dimensional frame of half dimensions imbedded in eight corner points of the cube and reach at their re-synthesis as four three dimensional frames of full dimensions.
7. These four re-synthesised three dimensional frames together with the fifth three dimensional frame at the centre shall be working out a solid dimensional frame of five dimensions of 5-space.
8. The transition from static cube to dynamic cube shall be leading to a creator space (4-space).
9. A step ahead, a transcendence from creator space (4-space) shall be leading to the transcendental worlds (5-space/𝔖).
10. It would be a very blissful exercise to chase this transcendental phenomenon of solid order.

◆ ◆ ◆

27
FOURTH VOWEL

1. Fourth vowel of *Devnagri* alphabet is *Rikara* (ऋ).
2. It being fourth of the row of vowels, as such it is associated with artifice 4/4-space/hyper cube 4.
3. This being, a step ahead of first three vowels, as such for its script format, one may first of all have a free look at the coordination of first three vowels along the script format of *Anubndh* of first *Maheshwara Sutra* namely letter (ण्).
4. 4-space is a creator space, and as such it provides four fold manifestation format for whole range of creations.
5. Accordingly 3-space/cube as 3-space domain, shall be of a manifestation features, of which 3-space itself would be in the role of domain fold (third fold) and 4-space shall be in the role of origin fold (fourth fold).
6. This as such shall be taking from domain fold (third fold/3-space/third vowel) to origin fold (fourth fold/4-space/fourth vowel) as an inward progression step.
7. This as such shall be helping us appreciate the script format of fourth vowel (ऋ).

8. One shall sit comfortably and have a fresh look at this script format, in continuity of the coordination format of first three vowels as first *Maheshwara* Sutra with (ण्) with as its *Anubandh.*
9. For full comprehension of the coordination format of first *Maheshwara* Sutra, one may have a look at the following expression for it.
10. Further, a step ahead one shall have a look at the following expression for second *Maheshwara* Sutra which coordinates fourth and fifth vowel with (क्) as its *anubandh.*

'4'

1. Artifice '4' is the first composite number, as much as that first three numbers namely 1, 2, 3 do not accept any divisor of value other than 1 and the number itself, so these are primes.
2. Artifice 4 is of a unique feature as much as that $4 = 2 + 2 = 2 \times 2$.
3. Still further, as it accept a pair of factors, as such it also absorbs the orientations because of addition and minus, as much as that $4 = 2 \times 2 = (-2) \times (-2)$.
4. These features are at the base of the creative features of 4-space as a creator space.
5. $4 = 1 + 1 + 1 + 1$ equips this whole number as an artifice of four features parallel to four dimensions.
6. However as $4 = 2 \times 2 = 2 + 2 = (-2) \times (-2)$, as such it uniquely splits the structural set-up of hyper cube 4 as representative regular body of 4-space as a set-up of a pair of two dimensional frames of half dimensions.
7. It is this feature of the split of the structural set-up in terms of a pair of half dimensional frames, which further makes 4-space being of unique features as that

its mathematics, science and technology would be workable in terms of half unit.

8. As such it would be a blissful exercise to chase 2 as 1.
9. Further it also would be a blissful exercise to chase 1 as 2.
10. Still further it would be a very blissful exercise to simultaneously chase 4-space as 1 as 2 and 2 as 1.

4-Space

1. 4-space is a creator space.
2. It accepts hyper cube 4 as its representative regular body.
3. The representative regular body of 4-space as well manifests at its four fold manifestation format.
4. It is this feature of the 4-space which deserves to be chased fully to imbibe its values completely.
5. It is in terms of this chase that one shall be glimpsing the features of idol of Lord Brahma, four head lord, the creator the supreme.
6. Lord Brahma, four head lord, creator the supreme manifests pair of eyes for his four heads and sits comfortably on a lotus seat of eight petals and meditates upon Lord Shiv, the Lord of transcendental worlds, within cavity of his heart and goes transcendental and multiplies ten fold as ten brahmas in Shiv Lok.
7. With this, sadkhas shall initiate themselves and meditates Brahma way and enlighten themselves about the fourth vowel, its script format and its composition split as (ऋ = र + इ).
8. A step ahead the fifth vowel (ऌ = ल् + ऱ् + इ), and the coordination of fourth and fifth vowels namely _ and ऌ as ऋऌक् as second *Maheshwara* Sutra deserves to

be chased fully for complete imbibing of its values, particularly as that the *Anubandh* is क्

9. It would further be an enlightenment as that the ancient wisdom approaches (क्) as first *varga* consonant, as *Anubandh* being of two fold ranges, firstly as Brahma, the presiding deity of 4-space/creators space and secondly as Shiv as presiding deity of 5-space/transcendental worlds.

10. It would further be an enlightenment as that letter 'क्' as of four folds is of values of letters 'र्, म्, ल्, ह्' and that the letter 'र्' is of the combined values of first three vowels as (,ʿ र) within spatial order of second vowel (ई)

Hyper Cube 4

1. Hyper cube 4 is the representative regular body of 4-space.

2. It is a manifestation layer of four folds parallel to artifices quadruple (2, 3, 4, 5)/(spatial dimension, solid boundary, hyper solid domain and transcendental origin).

3. These features are parallel to the features of idol of Lord Brahma, four head Lord, with pair of eyes in each head and Lotus seat being of eight petals.

   It would be blissful exercise to have comparative tabulation of the features of hyper cube 4 and idol of Lord Brahma as follows.

5. 4-space as origin fold of cube makes it centre a seat of hyper cube 4.

6. As such fourth vowel/4-space/hyper cube 4, as a third vowel/3-space/cube in a dynamic state, would help us appreciate the script format of fourth vowel (ऋ = र + इ)

Rikara as of a guided missile through 4-space.

7. Each component of the script format of fourth vowel, as such deserves to be chased to appreciate its feature and role.
8. It is infact a fourth degree curve trajectory.
9. One shall sit comfortably and permit the transcending mind to chase it as first part of second *Maheshwara* Sutra.
10. And at a next step, it to give place to second part of second *Maheshwara* Sutra with expression as follows.

Creative Order

1. 4-space in the role of dimension is a creative order of self-referral state (6-space).
2. It is in this role that 4th vowel deserves to be specifically chased.
3. With first *varga* consonant namely 'क'/*kakara* as *Anubanda* of second *Maheshwara* Sutra, the above chase would focus upon the sequential unfolding features of letter *Kakara* 'क', firstly as of set-up being 4-space and secondly as of set-up being 5-space.
4. It is in this context, the fourth vowel, itself as 4-space set-up, shall be focusing upon the spatial order.
5. Further it is in this context that the script format of fourth vowel (ऋ) shall be focusing upon the fourth degree curve as component of its script component.
6. It would be relevant to note that this fourth degree curve component, in its placement for vertically upward position stands accepted as a script format for first *Ushmana* letter.
7. It would further be relevant to note that first *Ushmana* letter is the dominant letter of the formulation (शिव)/ the lord of transcendental worlds (5-space).
8. One shall sit comfortably and permit the transcending

mind to chase these features of script format being availed.

9. Fourth degree curve in a dynamic state shall be acquiring additional features of added dimension which shall be providing transition from spatial order to solid order.
10. One shall sit comfortably and permit the transcending mind to remain in deep trans to fully comprehend and completely imbibe these values of fourth degree curve in dynamic state.

◆ ◆ ◆

28

FIFTH VOWEL

1. The script format of fifth vowel of *Devnagri* alphabet (ऌ) deserves to be chased as a sequential steps ahead of script format of fifth vowel.
2. This as such shall be transiting from (ऋ = र + इ) to (ऌ = ल् + र् + इ).
3. Here It would be relevant to note that while forth vowel ऋ = र+इ avails pairs of letters for its values and features, a step ahead, fifth vowel (ऌ = ल् + र् + इ) avails three letters for its values and features.
4. Further It would be relevant to note that while the prominent component of letter ऋ is र् , a step ahead the prominent component of letter ऌ are र् and ल.
5. These as such give a jump over letter 'म्'.
6. Still further these as of values of artifices '3 and 5'/3-space and 5-space respectively make it a feature of a shift from dimension (3-space) to domain (5-space).
7. This feature of the composition values of letter 'ऌ', as such shall be, making it of the features and values of 5-space.

8. Accordingly, the focus of fifth vowel would be upon 5-space within 4-space.
9. It is this feature of 5-space within 4-space, which infact is a step in continuity of internal progression, as of 4-space within 3-space and ahead, as of 5-space within 4-space.
10. This, this way, also would focus upon the transcendence phenomenon of coordination of dimension fold and domain fold and the same deserves to be chased thoroughly.

'5'

1. Artifice '5', like artifice 1, 2 and 3, does not accept a factor other than 1 and 5.
2. This, this way is uniquely of different features than that of artifice 4, which accepts 2 as its factor and the same is different than 1 and 4.
3. While within 4-space the feature $4 = 2 + 2 = 2 \times 2$ unifies indistinguishably for addition and multiplication operations but artifice $5 = 2 + 3$, while 2×3 is 6, greater than 5, and as such here in 5-space and addition and multiplication operations retain their distinctive features.
4. As such, the emergence of 5-space within 4-space, is a unique transcendental phenomenon which deserves to be chased like sky within space.
5. 5-space accepts 3-space as dimension and with it 5-space becomes is solid dimensional space.
6. And as 3-space itself is a linear dimensional set-up, as such the transcendental worlds (5-space) acquire sequential transcendental worlds (5-space) features as much as that 1-space leads to 3-space and ahead 3-space leads to 5-space.
7. On the other hand, 4-space accepts 2-space in the role of dimension.

8. And 2-space itself accepts 0-space in the role of dimension.
9. This, this way, with emergence of 5-space within 4-space, there emerge a super imposition of the structural set-up of transcendental steps (1, 3, 5) superimposed upon (0, 2, 4).
10. One shall sit comfortably and permit the transcending mind to chase 0 as 1 and 1 as 0 and to chase the transcendental phenomenon ahead of emergence of 5-space within 4-space.

5-Space

1. 5-space is a solid order space.
2. 5-space is also because of its transcendental values (1, 3, 5) is designated and is known as transcendental worlds (5-space).
3. Lord Shiv, five head lord with three eyes in each head has ten beautiful arms and is the presiding deity of the transcendental worlds (5-space).
4. Hyper cube 5 is the representative regular body of 5-space.
5. It would be blissful exercise to chase parallel features of idol of Lord Shiv and of hyper cube 5 within 4-space along its four fold manifestation format.
6. One shall chase five heads of the idol as five dimensions.
7. Further one shall chase three eyes in each head as a solid dimensional order.
8. Still further one shall chase ten beautiful arms of idol as ten boundary components.
9. Still further one shall chase origin of 4-space as transcendental seat.
10. And ahead one shall sit comfortably and permit the transcending mind to go Brahma way to multiply ten fold.

Hyper Cube 5

1. Hyper cube 5 accepts self-referral origin seat.
2. It is this feature of the origin fold of hyper cube 5, which deserves to be chased as a sequential step ahead of the transcendental origin of 4-space.
3. Self referral origin of 5-space, as a step ahead of transcendental origin of 4-space shall be taking ahead of ten fold multiplication of Lord Brahma as Ten brahmas to twelve fold multiplication of transcendental worlds as 12 sons of self-referral state of 6-space.
4. This phenomenon of internal progression and simultaneous multiplication of the domains at the boundary of the hyper cube of next order, deserves to be thoroughly chased.
5. It is this phenomenon of multiplication of domains which makes the manifestation within creator space (4-space) as of unique features.
6. With creator space (4-space) in the role of dimension of self-referral space accepting transcendental boundary makes the creative order of multifold features.
7. Ancient wisdom preserves the way this multifold creations deserve to be chased.
8. *Rig ved Samhita* simultaneously organises as of ten *mandals* and eight *austaks*.
9. It is like simultaneous chase of creative boundary of ten components of transcendental worlds and mundane boundary of eight components of the creator space (4-space).
10. One shall sit comfortably and permit the transcending mind to learn to chase the *Rigveda* way to chase multifold creative features.

Transcendental Order

1. Transcendental order (5-space) leads to unity state of pole star (7-space).

2. 5-space in the role of dimension, is its one role.
3. For complete comprehension of 5-space, one shall chase it in its different roles.
4. One set of roles for 5-space is as dimension, boundary, domain and origin.
5. The other sets of role for 5-space is as a transcendental base.
6. As 5-space itself accepts manifested four fold form as such 3-space, 4-space, 5-space and 6-space shall be contributing distinctive features.
7. Further as 3-space, 4-space and 6-space shall as well be having different roles to play, therefore the whole range of contributions for these spaces in their different roles are bound to make the structral set-up to be of very rich values and features.
8. One way to express these features is as 5 5 matrix format, as follows:

   1 2 3 4 5

   2 3 4 5 6

   3 4 5 6 7

   4 5 6 7 8

   5 6 7 8 9
9. This, this way brings into the contribution range the whole range of 1-space to 9-spaces.
10. This range of 1-space to 9-space is the range of first vowel to nineth vowel, and because of it, the first five vowels constitute a distinct class and these together get coordinated as (अक).

◆ ◆ ◆

29
SIXTH VOWEL

1. Sixth Vowel of *Devnagri* alphabet has script form (ऐ).
2. One shall have a fresh look at this script form of sixth vowel (ऐ) and chase its pair of opposite oriented curves.
3. Further it also be observed with a focus as that left curve is bigger and is having orientation from west to east.
4. While the other curve is smaller in size and is having orientation from east to west.
5. It is like a spiral like descending along the cylindrical valves of a well.
6. It is symbolic reach the water surface of the well and to have a dive deep for getting fulfilled.
7. Sixth vowel, as a step ahead of first five vowels is a step of transcendental decadence (6, 4, 2), as comparison to transcendental ascendance (1, 3, 5).
8. This, this way would be a transition from linear transcendental ascendance to spatial reach along transcendental decadence.
9. One way to look at is as that $2 + 3 = 5$ and $2 \times 3 = 6$.
10. Other way to look at is that first *Maheshwara* Sutra coordinates three vowels and parallel to it three dimensions

while the second *Maheshwara* Sutra coordinates pair of half dimensions of first dimension and parallel to it the coordination of fourth and sixth vowels, and a step ahead as third *Maheshwara* Sutra there being a coordination of pair of half dimensions of second dimension and parallel to it there being a coordination of sixth and seventh vowels and finally as fourth *Maheshwara* Sutra, there is a coordination of eighth and ninth vowels parallel to coordination of pair of half dimensions of third dimension.

'6'

1. The whole number 6 is perfect number, as much as that $1 + 2 + 3 = 1 \times 2 \times 3$.
2. Further 6 accept re-organization as $6 = 2 + 2 + 2$, which parallel to $3 = 1 + 1 + 1$ would be a mathematics of 2 as 1, as comparison to mathematics of 1 as 1.
3. It would be relevant to note that synthesis of three dimensions of all dimensional frames always lead to the same value and order of artifice 6, as synthesis of pair of dimensions leads to $N + N\text{-} (N - 2) = N + 2$ and a step ahead the synthesis of triple dimension shall be leading to $N + 2 + N - 2\ (N - 2) = 6$.
4. Parallel to values and order of artifice 6 is the set-up of 6-space.
5. Hyper cube 6, as representative regular body of 6-space, makes its feature as well parallel to artifice 6.
6. As six point lead to five units, as such parallel chase would be the coordination of artifices 6 and 5.
7. The internal coordination of the vertisis of a pentagon cut a pentagon frame around the centre/origin, and likewise would be the feature of hexagon.
8. It would be blissful exercise to chase artifice 5 and artifice 6, pentagon and hexagon, hyper cube 5 and hyper cube 6, 5-space and 6-space, and Triples (1, 3, 5) and (2, 4, 6).

9. It would be blissful exercise to have comparative chase for the idols of Lord Brahma, Lord Shiv and Lord Vishnu.
10. Further it would be a blissful exercise for the chase of triple artifices (9, 11, 13) and parallel to it (the 9 versions of hyper cube 4, 11 versions of hyper cube 5 and 13 versions of hyper cube 6).

6-Space

1. Ancient wisdom extensively preserves features values virtues and all that about 6-space.
2. It is Vishnu lok.
3. It is atman.
4. It is sun.
5. It is *Pursha* format.
6. It is a creative order space.
7. It accepts transcendental worlds (5-space) at its boundary.
8. Unity state is its source origin.
9. Its structures Asht Prakrati.
10. It as self-referral dimensional order leads to Nav Braham as origin source fold.

Hyper Cube 6

1. Hyper cube 6 is the representative regular body of 6-space.
2. Its structural features run parallel to idol of Lord Vishnu.
3. It has 13 versions parallel to 13 geometries of 6-space.
4. It works out a *sthapatya* measuring rod with Lord Vishnu as its presiding deity and Lord Brahma as the presiding deity of its measuring.

5. Its provide a format for self-referral state of consciousness.
6. It manifests self-referral values fulfilled within the transcendental domains from their origin.
7. Comprehension view of the set-up of hyper cube 6 can be had by chase of manifestation along 6   6 matrix format:

   1 2 3 4 5 6
   2 3 4 5 6 7
   3 4 5 6 7 8
   4 5 6 7 8 9
   5 6 7 8 9 10
   6 7 8 9 10 11

8. Of this format, the prominent role is of 4   4 matrix format

   3 4 5 6
   4 5 6 7
   5 6 7 8
   6 7 8 9

9. It would be blissful exercise to chase, firstly the manifestation layer (3, 4, 5, 6) and then the manifestation layer (6, 7, 8, 9).
10. The unity range (3, 4, 5, 6, 7, 8, 9) leads from Triloki to Nav Braham and it is this feature which deserves to be comprehended fully and imbibed completely.

Self-referral Order

1. 6-space in the role of dimension structures self-referral order in Asht Prakrati.
2. Synthesis of pair of 6-space as dimension leads to 8-space domain.

3. And a step ahead as synthesis of three such dimensions re-structure 6-space itself.
4. Synthesis of 4-such dimensions leads to 0-space.
5. It is this phenomenon of self-referral order which takes from 8-space to 6-space to 0-space which deserves to be chased fully.
6. With 6-space itself being of a creative order (4-space in the role of dimension), the pair of dimensions shall be leading to 6-space, the synthesis of 3-dimensions as well shall be leading to 6-space itself.
7. And a step ahead the synthesis of four dimensions (4-space in the role of dimension), shall be taking back to 4-space.
8. Therefore while the structures of 8-space would be as of values (6, 8, 6, 0), while the inner folds because of creative dimension of dimension of self-referral order shall be leading to the values (4, 6, 6, 4).
9. It would be a blissful exercise to sequentially chase (6, 8, 6, 0) and (4, 6, 6, 4) as (6, 4), (8, 6), (6, 6) and (0, 4).
10. One shall sit comfortably and permit the transcending mind to chase above structural features at dimension and at dimension of dimension level.

◆ ◆ ◆

30

SEVENTH VOWEL

1. Seventh vowel of as a script format (ओ).
2. One shall have a fresh look at the script form of (ओ).
3. It is of two parts first is as of first vowel, and second part is (ो).
4. The second part is again of two parts, namely (`) and (ा).
5. The first of these two, namely (`) is the *matra*/measure, value of sixth vowel.
6. The second of these two namely (ा) is the *matra* of first vowel, in its elongated form depiction (अ).
7. The composite form (ो) with *matra*/measure unit (ो) is a artifice 6 value flow/vowel sixth in flow/hyper cube 6 in dynamic state, i.e. of artifice value 7/(vowel 7).
8. This will help us comprehend the coordination of sixth and seventh vowels together as third *Maheshwara* Sutra.
9. One shall sit comfortably and permit the transcending mind to chase the script form and format of seventh vowel.

10. One shall permit the transcending mind to be in prolonged deep trans to fully comprehend and completely imbibe the values of the script form and format of seventh vowel (ओ).

'7'

1. Artifice '7' as unique feature.
2. It is the biggest prime numeral of ten place value system.
3. 3-space has seven geometries and corresponding to it cube has seven versions.
4. All the eight corner points of Cube permit coordination in terms of seven sequentially arranged edges of the cube.
5. Boundaries of hyper circles 1 to 7 are of sequential increase value while hyper circle 8 onwards are of sequential decrease value.
6. Unity state of consciousness is the seventh state of consciousness following waking state, dream state, deep sleep state, *turia* state, transcendental state and god state of consciousness.
7. *Atharav Ved* begins with the concept of *Trishapta* (3 and 7).
8. *Trishapata* (3 and 7) amongst other aspect is a *Triloki* (3-space/A) to pole star range.
9. It would be blissful exercise to reach at different coordination of 3 and 7, like 37, 37, 3 7, 3/7, 73 and so on like 3.7, 7.3 etc. etc.
10. It would be blissful exercise to chase the *Divya Ganga* flow of seven streams, 3 streams and 1 stream.

7-Space

1. 7-space is of pole star value.
2. Also it is a unity state values.
3. Within creator space (4-space) hyper cube 7 is its print out.
4. 7-space plays the role of origin of 6-space.

5. Parallel to it are the roles and coordination of pole star and Sun.
6. It would be blissful exercise to chase 7-space in its different role.
7. Of these the prominent roles are as dimensions, boundary, domain, origin, base, self-referral format and unity state.
8. Further 5-space as dimensional fold, 6-space as boundary fold, 8-space as origin fold and 9-space as base of origin fold are contributing different values and virtues to seven space.
9. The unique manifestation layer is (7, 8, 9, 10) with 7 + 8 + 9 + 10 = 32, as a life value.
10. The transcendental range (6, 7, 8, 9, 10) from Sun to Par Braham is a fire value of the existence phenomenon.

Unity State

1. Sadkhas aspire to attain unity state.
2. It is attained sequentially.
3. It is parallel to the chase of vowels 1 to 7.
4. Further parallel to hyper cubes 1 to 7.
5. Still further parallel to hyper circles 1 to 7.
6. It would be of the range of 1-space to 7-space.
7. It begins with waking state of consciousness and culminates into unity state of consciousness.
8. It covers the whole range of existence phenomenon with pole star as the origin of the solar universe.
9. It would be a life mission to be in a unity state.
10. It would be blissful to be in a unity state.

Unity Order

1. Unity order is the ascending order.

2. It is the optimum attainment of the existence phenomenon within human frame.
3. It attains Brahman range (9-space).
4. Unity order regulates itself of its own.
5. It regulates itself of its own as 9 = 32 and thereby linear order 3-space gets regulated by the solid spatial order.
6. It sustain its range as 9 = 32 and 8 = 23 fully limits the linear order existence phenomenon.
7. Sequentially one shall attain waking state and transition from waking state to dream state, and then after attainment of the dream state to have transition for deep sleep state and like that to sequentially progress.
8. Asht Prakrati is a step ahead of unity state and as such the ego of unity state cannot even reach near the Nature.
9. Further Braham is a step ahead of Asht Prakrati.
10. And Par Braham is even ahead of Braham.

◆ ◆ ◆

31

EIGHTH VOWEL

1. Eight vowel of *Devnagri* alphabet has script form (ऐ).
2. It *matra* / measure unit expression is.
3. One shall have a fresh look at and ऐ.
4. It would be observed as that both these frames are of pair of parts and each part of both fames is expression of the value and features of sixth vowel.
5. It would be relevant to note that 6-space in the role of dimension shall be synthesizing 8-space set-up as a pair of such dimensions as that the dimensional synthesis mathematics leads to N + N –(N – 2) = N + 2 which for N = 6 yields 6 + 2 = 8/8-space/8-space as 8-fold nature.
6. Fourth *Maheshwara* Sutra coordinates eight vowel and ninth vowel together and as such it would be blissful exercise to simultaneously chase artifice 8 and artifice 9.
7. Artifice 8 and artifice 9 accept re-organizations as a reflection pair (23 32).
8. It is this pairing as (23 32) which makes artifice 8 and 9/8-space and 9-space/vowels 8 and 9 as a class in them selves.

9. It is a self-contained class.
10. As such this pair of vowels deserve to be chased as such.

'8'

1. Artifice '8' has unique features.
2. It is a cube of the only even prime (2).
3. Cube has eight corner points.
4. Plane has eight directions (four directions and four sub directions).
5. Hyper cube 4 has eight solid boundary components.
6. Artifice 8 accepts re-organization (3, 5) parallel to solid dimensional order of transcendental worlds (5-space/$).
7. Hyper circle 8 onwards accept sequentially decreasing boundaries.
8. It would be blissful exercise to chase (1, 2, 3) as primes and (8, 9, 10) as composites.
9. It would be blissful exercise to chase the transcendental range (4, 5, 6, 7, 8).
10. Further it also would be a blissful exercise to chase eight space in different roles.

Asht Prakrati

1. Asht Prakrati is a step ahead of unity state.
2. One way to approach it is as 8-space.
3. Other way to approach it is as its printout in creator space (4-space) being of the format of hyper cube 8.
4. Ancient wisdom preserves that *Asht Prakrati* a structured by the self-referral order of Sun.
5. $1 + 2 + 3 + 4 + 5 + 6 + 7 + 8 = 36$ is the value of the reality format of existence phenomenon.
6. It would be blissful exercise to chase (1, 7), (2, 6), (3, 5) and (4, 4).
7. Parallel to it would be the chase (7, 1), (6, 2), (5, 3), and (4, 4).

8. Simultaneous chase would be as [(1, 7), (7,1)], [(6, 2), (2, 6)], [(5, 3) (3, 5)], [(4, 4), (4, 4)].
9. It would be very blissful exercise to chase [(4, 4), (4, 4)].
10. Ancient wisdom approaches *Asht Prakrati* as *Zad Prakrati* (mundane nature) and as *Chetan Prakrati* (consciousness nature).

Jad Prakrati

1. *Jad Prakrati* (mundane nature) is well demonstrated by the functional phenomenon of 'roots' of trees.
2. It is a mechanical operational system within well framed functional domain.
3. It is more or less a phenomenon of automation.
4. It though may be optimum functional efficiency but is to remain upon the year marked format.
5. There is determinism in it.
6. It has no operational discretion.
7. It has its demarcated limitations in its dimensional order (6-space).
8. It is this demarcated limitations which is inherent in the dimensional features in terms of which the dimensional frame splits into a pair of dimensional frames of half dimensions.
9. It is in terms of this that the transcendental decadence leads to spatial order of the values $2 \times 4 \times 6 = 48$, as the value of tree.
10. It would be a blissful exercise to chase the phenomenon of *Jad Prakrati* operationally functional in the existence phenomenon of Trees.

Chetan Prakrati

1. *Chetan Prakrati* (consciousness nature) on the other hand is regulated by the Brahman foundation fountaining from the origin.

2. It as such has a shift from 6-space as dimension of 8-space to 9-space as origin of 8-space.
3. Here the reflection pairing as (23, 32) has its role to play which adds conscious discretion as shift from base to index inherently imbedded in the set-up.
4. In *Jad Prakrati* (mundane nature) the focus vacillated between $(2 + 2 + 2)$ to $(2\times2\times2)$.
5. As such it would be a blissful exercise to chase firstly $(2 + 2 +2)$ and then $(2\times2\times2)$ to comprehend and imbibe the values of *Jad Prakrati* (mundane nature).
6. Still further it would be a very blissful exercise to firstly chase 23 and then 32 to comprehend and imbibe the values of *chetan prakrati* (consciousness nature).
7. One shall sit comfortably and permit the transcending mind to be face to face with the distinguishing feature of $2 + 2 + 2$ from $2\times2\times2$ set-up.
8. Further one shall sit comfortably and permit the transcending mind to be face to face with the distinguishing feature of 23 and 32 set-ups.
9. One shall permit the transcending mind to remain in prolonged sittings of deep trans to fully comprehend and to completely imbibe the differentiating features of *Jad Prakrati* and *chetan Prakrati.*
10. It would further be a blissful exercise to permit the transcending mind to chase the way *Brahman* values get fulfilled with an *Asht Prakrati* to make it a *Chetan Prakrati.*

◆ ◆ ◆

32

NINTH VOWEL

1. Ninth vowel of *Devnagri* alphabet as a script frame ''.
2. It is a format of 8-space in a dynamic state.
3. 8-space has a degree of freedom of motion in 9-space.
4. The transition from artifice 8 to 9 is parallel to the transition from the set-up of eight corner point of cube to eight corner points coordinated with the centre as the ninth point.
5. 9 is the largest numeral and parallel to it 9-space is the largest space.
6. It is designated and known as *Nav Braham.*
7. *Nav means* '9' and it also means new.
8. It is new to *Asht Prakrati.*
9. It would be relevant to note that transcendental worlds (5-space) as well are new to the creative space.
10. It would be blissful exercise to chase re-organization of 9 = 4 +5 as the transcendental worlds enveloped within creative space.

'9'

1. 9 is the largest numeral.
2. Cube has a nine point fixation.

3. Hyper cube 4 has 9 versions parallel to nine geometries of 4-space.
4. (01, 10) constitute a reflection pair and 9 = 10 – 01.
5. 6-space as dimension leads to 9-space as origin.
6. 7-space plays the role of dimension of 9-space.
7. 8-space envelops 9-space.
8. 9-space plays the role of origin of 8-space.
9. *Nav Braham* envelops *par Braham.*
10. Sadkhas remain ever blissful with their unity state fulfilling them *Brahaman* values and also of Par *braham.*

Nav Braham

1. *Nav Braham* is 9-space.
2. It is a new space.
3. It is ahead of *Asht Prakrati; Jad Prakrati* as well as *chetan Prakrati.*
4. Transcendental worlds (5-space) because of transcendence range (5, 6, 7, 8, 9) are ever new, may it be in dimensional order or in its attainment as Brahman source origin.

Braham

1. *Braham* is ahead of creations.
2. It is like *atman* vis-à-vis body.
3. It is like 4-space vis-à-vis 3-space.
4. It is like *Nav Braham* vis-à-vis *Asht Prakrati.*

Par Braham

1. *Par Braham* is *Braham* and is also different than Braham.
2. *Par Braham* is there in whole range of *Nav Braham*, and is still different than them all.
3. All this what ever it is, it is not *Par Braham.*
4. *Par Braham* is *Par Braham.*

◆ ◆ ◆

SECTION – 3: SPECIAL FEATURES

33. Existence Phenomenon
34. Five Basic Elements
35. Sun to Earth Range
36. Shad Chakra Format
37. Transcendental State
38. First Basic Element
39. Second Basic Element
40. Third Basic Element
41. Fourth Basic Element
42. Fifth Basic Element
43. Sun to Earth Eange
44. Pole Star to Water Range
45. Asht Prakrati to Fire Range
46. Nav Braham to Air Range
47. Par Braham to Space Range
48. Transcendental Carriers-1
49. Transcendental Carriers-2
50. Transcendental Carriers-3
51. Transcendental Carriers-4
52. Transcendental Carriers-5

33
EXISTENCE PHENOMENON

1. One shall sit comfortably and permit the transcending mind to chase its existence in reference to one's body.
2. One may first have a pause to focus upon one's head.
3. Then one shall have a focus upon one's brain.
4. One shall again have a pause and focus upon one's 'heart'.
5. Then upon one's senses one by one.
6. One shall once again have a pause and have a focus upon one's pulse.
7. Then upon one's breathing.
8. It would be a blissful experience to chase the way the sensory impulses reach one's mind.
9. And the way mind transcends from the sensory domain to the intelligence field and ahead into consciousness core,
10. And the way one's transcending mind brings one face to face with the existence phenomenon.

◆ ◆ ◆

34
FIVE BASIC ELEMENTS

1. One shall sit comfortably and permit the transcending mind to chase the values and order of five basic elements (Earth, Water, Fire, Air, Space).
2. These elements values deserve to be in their sequence and order of these elements as no. 1 Earth, 2, Water, 3. Fire, 4. Air and 5. Space.
3. Sequentially one shall focus upon the values of these elements in the manifestation of human frame.
4. One way to approach the values of these five elements in the context of the human frame is in reference to the solid content/bones, liquids, temperature, breathing air and the space within the human frame.
5. The other way to approach the values of these five basic elements, in the context of human frame is in terms of five senses.
6. One shall sit comfortably and permit the transcending mind to simultaneously chase the values and order of first element and first sense.

7. Second shall sit comfortably and permit the transcending mind to simultaneously chase the values and order of first element and first sense.
8. Third shall sit comfortably and permit the transcending mind to simultaneously chase the values and order of first element and first sense.
9. Fourth shall sit comfortably and permit the transcending mind to simultaneously chase the values and order of first element and first sense.
10. It would be blissful exercise to simultaneously chase values of elements and senses in the sequence and order of the elements and senses.

◆ ◆ ◆

35

SUN TO EARTH RANGE

1. One way to approach the existence phenomenon is as 'Sun to Earth range'.
2. 'Sun to earth range', is a six steps long range : Sun, Space, Air, Fire, Water and Earth.
3. 'Sun is of the order and values' of 'atman'/ soul.
4. It would be blissful exercise to chase this six steps long range for Sun to Earth parallel to six shad chakras/ eternal circuits format of human body.
5. One way to approach these parallel ranges in the sequential order of self-referral order, transcendental order, creative order, solid order, spatial order and linear order.
6. Self-referral order is of the values of 6-space in the role of dimension.
7. Transcendental order is of the values of 5-space in the role of dimension.
8. Creative order is of the values of 4-space in the role of dimension.

9. Solid order is of the values of 3-space in the role of dimension.
10. Spatial order is of the values of 2-space in the role of dimension and linear order is of the values of 1-space in the role of dimension.

◆ ◆ ◆

36

SHAD CHAKRA FORMAT

1. One shall sit comfortably and permit the transcending mind to chase *shad chakra* formats as hyper cubes format.
2. It would be blissful exercise to chase first chakra as of the format of hyper cube 1/interval/(–1, 0, 1, 2).
3. Second *chakra* format deserves to be chased as of the format cube 2/ square (0, 1, 2, 3)
4. Third *chakra* format deserves to be chased as of the format hyper cube 3/ (1, 2, 3, 4)
5. Fourth *chakra* format deserves to be chased as of the format hyper cube 4/ (2, 3, 4, 5)
6. Fifth *chakra* format deserves to be chased as of the format hyper cube 5/ (3, 4, 5, 6)
7. Sixth *chakra* format deserves to be chased as of the format hyper cube 7/ cube (4, 5, 6, 7)
8. Fourth *chakra* is of the format of idol of Lord Brahma.
9. Fifth *chakra* is of the format of idol of Lord Shiv.
10. Sixth *chakra* is of the format of idol of Lord Vishnu.

◆ ◆ ◆

37

TRANSCENDENTAL STATE

1. Sixth *chakra* is located at the top of the head.
2. It is of the format of idol seat of Lord Vishnu.
3. Being of the format of idol of Lord Vishnu, it straight a way gets coordinated with orb of the sun through sun light.
4. It is this coordination of *Braham Randra* (sixth chakra) with Sun which creates a transcendental state.
5. The transcendental state is state, is the state of 5-space in the role of dimension.
6. It is in this state that 'being' rides the transcendental carriers within the rays of the Sun.
7. It is in this state that one comes face to face with one self.
8. It is the state in which the Being leaves the body and carries mind in transcending state along with it through rays of the Sun for its destination in the Sun.

9. One shall sit comfortably and permit the transcending mind to remain in prolonged deep trans to create transcendental state.
10. And further and to state its eternal seat at the transcendental boundary of self-referral orb of the Sun.

◆ ◆ ◆

38

FIRST BASIC ELEMENT

1. The existence phenomenon within human frame attain its transcendental state with the coordination of the *Braham Randra* (sixth chakra) with orb of the Sun through rays of the Sun.
2. In this state the creative order (4-space in the role of dimension of 6-space) of *Braham Randra*/sixth chakra in this transcendental state, transits and transforms into a transcendental order (5-space in the role of dimension).
3. It is this transition and transformation of creative order into transcendental order, in a transcendental state, as such sequentially transforms five basic elements for their features and values.
4. The earth element, being the first basic element, is of a linear order in creator space (4-space).
5. With transition and transformation of the creator space (4-space) into features and values of the transcendental space, the values of the first basic element, as such, accordingly transform.

6. Creator space (4-space) during this transition and transformation, transits from its role of domain fold into boundary fold of transcendental worlds (5-space).
7. With it the solid boundary of 4-space transits and transforms into solid boundary of the boundary of the transcendental space (5-space).
8. One shall sit comfortably and permit the transcending mind to chase this transition and transformation of the features and values of solid boundary/3-space into solid boundary of the boundary of 5-space.
9. It would be relevant to note that Earth/Solids/3-space/linear order in the process transforms into the solid order of 5-space.
10. It would be blissful exercise to chase these features of transition and transformation of first basic element/earth/linear order set-up, transiting and transforming into a solid dimensional order set-up of 5-space.

39

SECOND BASIC ELEMENT

1. Water, the second basic element is a set-up of a spatial order.
2. Being a spatial order, it is the order of the creator space (4-space).
3. During transition and transformation of the creative order in transcendental state, the features and values transits and transforms from the manifestation set-up into transcendence set-up.
4. It is this phenomenon of transition and transformation from manifestation set-up into transcendence set-up that the spatial order shall be transiting and transforming into its new role of dimension of dimension of 6-space.
5. One shall sit comfortably and permit the transcending mind to chase this transcendence phenomenon in which a space in a dimensional role transits and transforms into a dimension of dimension role.
6. It would be relevant to note that, it is in this process that the first basic element/Earth/linear order had

ultimately transited into the role of solid order of 5-space.

7. Likewise, here the second basic element/water/spatial order set-up is attainting transition and transformation of transcendental order and taking from 2-space to 4-space to 6-space.
8. It would be relevant to note that as this is essentially the phenomenon of 4-space where in parallel to $4 = 2 + 2 = 2\times2 = -2\times-2$, firstly the addition and multiplication operations gets superimposed and secondly the orientations as well get super imposed.
9. Still further as 4-space as nine geometries as such the transcendence from above as well as from below i.e. from 9 to 7 to 5 and 1 to 3 to 5 reach the transcendental middle.
10. As such one shall revisit the first element in creator space (4-space) and its transition and transformation of features and values of transcendental state which shall be taking from the two fold coverage as a single range 1 to 3 to 5 to 7, and the spatial order characteristics would make it a phenomenon of 1 to 3 to 7 to 9.

◆ ◆ ◆

40
THIRD BASIC ELEMENT

1. One shall revisit the sequential range 1 to 3 and 9 to 7 as 1 to 3 to 7 to 9, with a jump of four steps at the middle from 3 to 7.
2. This jump from 3 to 7, deserves to be chased and reaches to imbibe its full comprehension.
3. It would be relevant to note that 3-space has seven geometries.
4. Further it would be relevant to note that this four steps gap is like accepting bridging in terms of artifice 4/4-space.
5. A step ahead, at the phase and stage of transition and transformation of the features and values of third element/fire/solid order set-up shall be providing bridging of six steps gap.
6. This shall be the coverage of the middle of six steps of the range 3 to 5 and 11 to 13.
7. One shall sit comfortably and permit the transcending mind to sequentially chase the first transcendental range (1, 3, 7, 9) and (3, 5, 11, 13).

8. It would be relevant to note that six steps long gap at the middle is parallel to six boundary components of cube/3-space/solid order.
9. Further It would be relevant to note that artifice 3 is parallel to three geometries of 1-space, artifice 5 being parallel to five geometries of 2-space, artifice 11 being parallel to eleven geometries of 5-space and artifice 13 being parallel to 13 geometries of 6-space.
10. One shall sit comfortably and permit the transcending mind to sequentially chase the transition and transformation of features of first second and third element during transcendental state.

◆ ◆ ◆

41

FOURTH BASIC ELEMENT

1. Air is the fourth element.
2. Hyper cube 4 is the representative regular body of 4-space.
3. The domain boundary ratio of hyper cube 4 is A4: 8A3.
4. It would be blissful exercise to chase bridging of eight steps gap at the middle of the transcendental range (5, 7, 15, 17).
5. It would be relevant to note that this quadruple is parallel to the quadruple of geometries of (2-space, 3-space, 7-space and 8-space).
6. It would be blissful exercise to sequentially chase the transition and transformation of the features and values of fourth element during transition state.
7. One shall sit comfortably and permit the transcending mind to have a comparative chase of the features and values of first element and fourth element.

8. Likewise one shall chase simultaneously the features and values of second element and fourth element.
9. Further to chase simultaneously the features and values of third element and fourth element.
10. It further would be blissful exercise to simultaneously chase the features of all the first four elements together in transcendental state.

◆ ◆ ◆

42

FIFTH BASIC ELEMENT

1. Hyper cube 5 is the representative regular body of 5-space/transcendental worlds.
2. The domain boundary ratio of hyper cube 5 A5: 10B4.
3. It would be blissful exercise to chase bridging of ten steps long gap at the middle of the transcendental range (7, 9, 19, 21).
4. One shall sit comfortably and permit the transcending mind to chase the way Lord Brahma, creator the Supreme, attains transcendental state.
5. During transcendental state, Lord Brahma multiplies 10 fold.
6. Ten *brahamas* constitute the creative boundary for the transcendental worlds.
7. A step beyond is a transcendental step.
8. It is a transcendental step of Being riding the transcendental carriers within rays of the sun.
9. The attainment stage for the Being within orb of the Sun is of order and value of 12 steps long gap at the middle

of the transcendental range being bridged by *Dwadash Adityas*.

10. One shall sit comfortably and permit the transcending mind to remain in prolonged deep trans to be face to face with the transcendental phenomenon within rays of the Sun attaining transcendental range (9, 11, 23, 25) where $9 = 3 \times 3$ and $25 = 5 \times 5$ well indicate the spatial attainment and $11 = 2 \times 5+1$ and $23 = 2 \times 11 + 1$ and artifice 23 being of the order and values of the syllables of *Gyatri mantra* enlighten about the methodology and technology of the transcendental carriers.

◆ ◆ ◆

43

SUN TO EARTH RANGE

1. One shall sit comfortably and permit the transcending mind to and to chase Sun to Earth range as of six steps range viz. (1) Sun, (2) Space (3) Air (4) Fire (5) Water and (6) Earth.
2. This range is in a sequential decreasing dimensional order, as much as that Sun is of the values and order of 6-space, space is of the order and value of 5-space, Air is the order and values of 4-space, Fire is of order and value of 3-space, Water is of the order and value of 2-space and Earth is of the order and value of 1-space.
3. This range is as such of the format of hyper cube 6, hyper cube 5, hyper cube 4, hyper cube 3, hyper cube 2 and hyper cube 1.
4. These formats are of domain boundary ratio AN: 2N BN-1 for N=6, 5, 4, 3, 2, 1.
5. The boundaries of this formats range are of sequential values parallel to artifices (12, 10, 8, 6, 4, 2).

6. The domains of this formats range are of the sequential value parallel to the artifices values (6, 5, 4, 3, 2, 1).
7. The transcendental measuring rod within 6-space/hyper cube 6 is constituted by hyper cubes 0 to 5 as of geometric values N× 2 + 1, for N = 0, 1, 2, 3, 4, 5, i.e. 1, 3, 5, 7, 9, 11.
8. This way the boundary values (2, 4, 6, 8, 10, 12) together with geometric values (1, 3, 5, 7, 9, 11) constitute a coverage for Sun to Earth range.
9. This split as (2, 4, 6, 8, 10, 12) as from 2 to 12/B to L, and as (1, 3, 5, 7, 9, 11) as from 1 to 3 to 11/A to C to K, makes a BLACK range.
10. It would be blissful exercise to revisit this range to chase the way Sun light makes this BLACK range into WHITE range; W-hit-E i.e. 23-hit (37) – 5 i.e. *Gyatri* to hit the Being to make it transcendental/*Gyatri mantra* to transcendentally approach the whole range of existence phenomenon as the range of Mantras within transcendental domain.

◆ ◆ ◆

44

POLE STAR TO WATER RANGE

1. Parallel to 'Sun to Earth' range deserves to be chased 'Pole Star to Water' range.
2. The Sun to Earth range is of the values and order of sequential range (6, 5, 4, 3, 2, 1).
3. Parallel to it, Pole star to water range, is of the values and order of the sequential range (7, 6, 5, 4, 3, 2).
4. Like Sun to Earth range, Pole Star to water range as well is of self-referral values format but superimposed upon the transcendental values.
5. These values formats are of the sequential format of 6 6 matrix formats:

6	5	4	3	2	1
7	6	5	4	3	2
8	7	6	5	4	3
9	8	7	6	5	4
10	9	8	7	6	5
11	10	9	8	7	6

6. It would be a blissful exercise to comprehend and imbibe the transcendental values of this phenomenon of Sun to Earth range as of artifices values (6, 5, 4, 3, 2, 1) transiting and transforming into Par Braham to Space range as of artifices values (11, 10, 9, 8, 7, 6).
7. This attainment is going to be for *Ekadash Rudras*/11 geometries of 5-space/11 versions of hyper cube 5/geometric values of the transcendental boundary of Sun/artifice 11 = 5 + 6, as of the order of 'Sun'/ artifice 6.
8. Here It would be relevant to note that the transcendental boundary of Sun/6-space/hyper cube 6 is of 12 components.
9. These 12 transcendental components of the transcendental boundary of the Sun, as such transit and transform as *dwadash adityas*/12 suns.
10. It would be blissful exercise to chase this phenomenon time and again to fully comprehend and imbibe the values of the transcendental phenomenon of fulfilling the transcendental domains with self-referral values and the transcendental domains transforming and multiplying as 12 transcendental domains, parallel to Lord Brahma transforming and multiplying as 10 *Brahamas* within transcendental domains.

◆ ◆ ◆

45

ASHT PRAKRATI TO FIRE RANGE

1. One shall have a fresh look at 6×6 matrix format:

6	5	4	3	2	1
7	6	5	4	3	2
8	7	6	5	4	3
9	8	7	6	5	4
10	9	8	7	6	5
11	10	9	8	7	6

2. Asht Prakrati to fire range is of the order and values of the third row of the above format: 8, 7, 6, 5, 4, 3.
3. It would be blissful exercise to chase this self-referral range firstly as the range of dimensional spaces i.e. (8-space, 7-space, 6-space, 5-space, 4-space, 3-space).
4. Then, this range shall be chased in terms of their respective representative bodies/hyper cubes within creator space (4-space) namely (hyper cube 8, hyper cube 7, hyper cube 6, hyper cube 5, hyper cube 4, hyper cube 3).

5. Then this hyper cubes range shall be approached in terms of the transcendental values fulfilled within the domains of this range of hyper cubes, namely hyper cube 8 to be approached in terms of the transcendental values fulfilled within 8-space domain and like that to be approached the other hyper cubes of this range as well.
6. Then, a step ahead this sequential range deserves to be approached as self-referral values superimposed upon the transcendental values fulfilled within the domains of these hyper cubes.
7. A step ahead, these are to be approached as unity state features and values for the domains of such hyper cubes.
8. It, as such would make a phase and stage for transition and transformation from the unity state to the Asht Prakrati phase and stage of values and order.
9. One shall sit comfortably and permit the transcending mind to remain in prolonged deep trans to be phase to phase with the transcendental phenomenon of transition and transformation from unity state to the Asht Prakrati values and order.
10. It would be blissful exercise to chase this phenomenon as the phenomenon of transition from artifice 7 to artifice 8/hyper cube 7 to hyper cube 8/7-space to 8-space/ pole star to eight fold nature/from phenomenon of increasing nature of hyper circles 1 to 7 to decreasing nature of hyper circle 8 onwards/from the sequential order (1, 2, 3, 4) to (1, 2, 3, 8)/from *Triloki* to *Trimurti*.

◆ ◆ ◆

46

NAV BRAHAM TO AIR RANGE

1. *Nav Braham* to Air range is of sequential range values (9, 8, 7, 6, 5, 4).
2. Artifice 9 accepts re-organization as 3 3, i.e. of a square format.
3. Artifice 8 accepts re-organization as 2 2 2, i.e. of a cube format.
4. One shall sit comfortably and permit the transcending mind to fully comprehend and imbibe the phenomenon of transition and transformation from features and values of 32 to 23 , as a base-index reversion pairing (32 and 23).
5. It would be relevant to note that the creative dimensional order/4-space in the role of dimension creates manifestation of folds (4, 5, 6, 7) with 8-space as base of the origin fold and 9-space as the self-referral format for the transcendental base of the origin, which embeds geometric values in 4-space as of 9 geometries and parallel to it there being 9 versions of hyper

cube 4 itself being the representative regular body of 4-space.

6. Further One shall sit comfortably and permit the transcending mind to fully comprehend and imbibe the phenomenon of transition and transformation from features and values of artifice 8 to artifice 7/8-space to 7-space in relevance to 9-space.
7. Further One shall sit comfortably and permit the transcending mind to chase the transition and transformation from artifice 7 to artifice 6/7-space to 6-space, in relevance to 9-space as self-referral format for the transcendental base.
8. Further One shall sit comfortably and permit the transcending mind to chase the transition and transformation from artifice 6 to artifice 5/6-space to 5-space, in relevance to 9-space as self-referral format for the transcendental base.
9. Further One shall sit comfortably and permit the transcending mind to chase the transition and transformation from artifice 5 to artifice 4/5-space to 4-space, in relevance to 9-space as self-referral format for the transcendental base.
10. Further One shall sit comfortably and permit the transcending mind to chase the transition and transformation from artifice 4 to artifice 3/4-space to 3-space, in relevance to 9-space as self-referral format for the transcendental base.

◆ ◆ ◆

47

PAR BRAHAM TO SPACE RANGE

1. *Par Braham* to space range is the ultimate attainment range.
2. It is the range of comprehension domain of the transcendental mind.
3. *Sadkhas* fulfilled with intensity of urge to comprehend and imbibe the values of this range shall purify their sensory domain.
4. Further they shall fulfil their intelligence domain with transcendental values.
5. Still further they shall continuously remain in prolonged deep trans to intensify the urge to fulfil their consciousness field with virtues of the transcendental worlds.

◆ ◆ ◆

48

TRANSCENDENTAL CARRIERS-1

1. Ancient wisdom enlightens as that our *Triloki* (त्रिलोकी) flourishes within *Jyoti* (ज्योति)/*Jyotirmadye Triloki Manormam*.
2. Ancient wisdom further enlightens as that *Jyoti* flourishes within Rays of the Sun as *Ativahkas*/ transcendental carriers.
3. One shall sit comfortably and permit the transcending mind to chase the sensory messages, intelligence bits, consciousness impulses, transcendental bliss and *Brahaman* fulfilness.
4. Further, One shall sit comfortably and permit the transcending mind to chase the features and values of the ultimate four fold supports domains of '*Manas, Budhi, Chit* and *Ahankar*'.
5. Further One shall sit comfortably and permit the transcending mind to chase the 101 nerves as support basis of ultimate supports domain of '*Manas, Budhi, Chit* and *Ahankar*' and the 101th Nerve, the *Sushmana Nadi* coordinating the Shad Chakras of the human frame.

6. Still further One shall sit comfortably and permit the transcending mind to chase the way *Sushmana Nadi* coordinates the Shad Chakras of the human frame with the origin core of the *sixth chakra/Braham Randra*.
7. Ancient wisdom enlightens as that the Being through the *Sushmana Nadi* reaches *Braham Randra*.
8. Reaching *Braham Randra*, the Being rides the transcendental carriers of Sun light.
9. Ancient wisdom enlightens that the take off of Being from *Braham Randra* is as per the features and order of rays of the Sun (रष्म्यानुसारी).
10. *Sadkhas* fulfilled with the intensity of urge to chase the Being riding the transcendental carriers shall continuously permit the transcending mind to be fulfilled with transcendental bliss by remaining in prolonged deep trans time and again.

◆ ◆ ◆

49

TRANSCENDENTAL CARRIERS-2

1. The simultaneous flourishing of *Triloki* for manifesting human frame for the existence phenomenon of Being within human frame and *Jyoti* playing the role of transcendental carriers for the take off of the Being from the *Braham Randra,* is the unique transcendental phenomenon which deserves to be chased by the *sadkhas*.
2. Ancient wisdom enlightens us as that Lord Shiv, is the presiding deity of this transcendental phenomenon.
3. Lord Shiv is the five head Lord with three eyes in each head.
4. The features and order of the idol of Lord Shiv manifest the format of hyper cube 5, a solid order set-up.
5. One shall sit comfortably and permit the transcending mind to chase the solid order dimensional frame of the transcendental worlds presided by Lord Shiv.
6. The sequential chase of this five dimensional frame under the synthesis process would lead to the sequential

parabolic range of values of artifices (3, 5, 6, 6, 5, 3).

7. One shall sit comfortably and permit the transcending mind to be face to face with this transcendental phenomenon of the solid order transcendental range of parabolic features, with its ascending range taking the Being in three steps of artifices values (3, 5, 6) from within the *Braham Randra* to the orb of the Sun.
8. The *Braham Randra* being the seat of hyper dimensional order (4-space in the role of dimension) set-up of self-referral space (6-space as domain), as such the center/core/origin of this seat shall be of transcendental order and values (of 5-space), which being of solid order, shall be providing a take off for the Being.
9. It would be blissful exercise to sequentially chase the transcendental progression of the pilgrimage of Being riding the transcendental carriers. With this take off position, as being of values of artifice 3/(3-space in the role of dimension), which at its next step of availability of second solid dimension shall be synthesizing a transcendence path of artifice value 5, as the synthesis of pair of solid dimensions shall be of such order and value as 3 + 3 – 1 = 5/solid dimension value + solid dimension value – linear dimension of solid dimension value.
10. It would further be blissful exercise to be face to face with the attainment phase and stage of Being riding the transcendental carriers reaching orb of the Sun/6-space/artifice value 6 as synthesis value of three solid dimensions/5 + 3 – 2 = 6/synthesis value of first two solid dimensions (5) + third solid dimension value (3) – pair of linear dimensions of solid dimension (2) = total value 6.

◆ ◆ ◆

50

TRANSCENDENTAL CARRIERS-3

1. The ascending limb (3, 5, 6) of parabolic range (3, 5, 6, 6, 5, 3) emanates from the *Braham Randra* where is the seat of Shad Chakra (6-space) of hyper dimensional order.
2. The origin of 6-space being 5-space as such it is a seat of transcendental order in terms of which transcendence takes place from domain to dimension of dimension.
3. Parallel to the transcendence from domain to dimension of dimension takes place ascendance from dimension of dimension to domain because of the spatial order of creator space (4-space) within which not only the addition and multiplication get superimposed but also the orientations as well gets superimposed.
4. Moreover it is because of the features and order of the creator space (4-space), a manifestation format becomes available on which four consecutive dimensional spaces content manifest a four fold manifestation layer.
5. It is on this format that (0, 1, 2, 3) manifestation layer emanates at the seat of *Braham Randra*, and thereby the

emergence and availability of a solid dimension.

6. It is this phenomenon which within spatial order creator space (4-space) splits in three dimensional frame into a pair of three dimensional frames of half dimensions with their origin getting superimposed by a *swastik* frame/spatial order dimensional frame of four spatial dimension.
7. It is the *swastik* churning of the origin coordinating the pair of three dimensional frames of half dimensions that transcendental values and order get fulfilled within creator space (4-space).
8. A step ahead of this attainment of emergence of superimposition of transcendental values and features within creator space (4-space), that self-referral values come into play as of the order and value of 6-space of hyper dimensional (4-space) order.
9. It would be a very blissful exercise to permit the transcending mind to chase this phenomenon of sequential progression of ascendance along the parabolic limb (0, 3, 5, 6) as the intelligence embedded transcendental carriers path for the pilgrimage of Being after its take off from the *Braham Randra* for the self-referral attainment of the order and values of orb of the Sun.
10. It would further be blissful exercise to chase the phenomenon of Being transcending from orb of the Sun to *Braham Randra* of human frame along the second limb (6, 5, 4, 3, 0) of the parabolic range (0, 3, 5, 6, 6, 5, 3, 0).

◆ ◆ ◆

51
TRANSCENDENTAL CARRIERS-4

1. Ancient wisdom enlightens that being after attaining self-referral position within orb of the Sun, has two fold paths, firstly as to ascend ahead to *Brahaman* domain or to descend again for existence within human frame.
2. It is the feature of the upper joint (6, 6) of the parabolic range, which shall be approached for its values and features.
3. It is a self-referral state.
4. Scriptures approach the state as a split of self-referral sphere into a pair of hemispheres/northern hemisphere and southern hemisphere/*Uterrayana* and *Dakishyana*.
5. These are the paths of liberation and of continuation in life death cycles.
6. Ancient wisdom focuses upon them and one illustrative and one scriptural illustrative position preserved is that of *Mahabharata* warrior *Bhishma* while having fallen down with injuries in the battle field had under is strong will waited for the upper tune moments of the

Utterranyana time for is last breath so that his being gets liberated from the life death life-cycle.

7. *Jyoti* flows as a *Divya Ganga* of seventh streams parallel to the formats of seven geometries of three-space/seven versions of cube, as of signatures range (–3, –2, –1, 0, 1, 2, 3).
8. It is this signature ranges which is parallel to the transcendental carriers path emanating from the *Braham Randra* with 0 signature format superimposed upon the core/origin seat of *Braham Randra*.
9. It would be blissful exercise to chase the next milestone along the transcendental carriers path which shall be taking to sky (5-space) within space (4-space).
10. Further it would be a very blissful exercise to chase the final phase and stage of attainment along the transcendental carriers path ending in orb of the Sun for a self-referral pause.

◆ ◆ ◆

52
TRANSCENDENTAL CARRIERS-5

1. From *Braham Randra* orb of the Sun is the first phase and attainment of the transcendental carriers within orb of the Sun.
2. It is the first pause of the being.
3. It is the self-referral pause of the pilgrimage of Being.
4. Ahead is the *Brahaman* phase of pilgrimage of Being.
5. This is the pilgrimage phase of attaining liberation from life death cycle journey.
6. Ancient wisdom enlightens us about this transcendental phenomenon as of three steps: (1) 0, 1, 2, 3 (2) 3, 4, 5, 6 and (3) 6, 7, 8, 9 manifested layers coverage of the *Brahman* range.
7. These three steps may be designated as (1) reaching uptill *Braham Randra* (2) Attaining orb of the Sun/self-referral state and (3) Ahead on the path of Brahman pilgrimage.

8. One shall sit comfortably and permit the transcending mind to chase first step of manifestation layer (0, 1, 2, 3) as *Triloki* existence phenomenon.
9. Further One shall sit comfortably and permit the transcending mind to chase second step of manifestation layer (3, 4, 5, 6) as *Trimurti* existence phenomenon.
10. And, still further, One shall sit comfortably and permit the transcending mind to chase third step of manifestation layer (6, 7, 8, 9) as *Brahaman* sustenance of existence phenomenon.

◆ ◆ ◆

SECTION – 4: KNOWLEDGE SYSTEMS OF GANITA SUTRAS

53. Pure Knowledge: Self-Organizing Value
54. Compactified Consciousness States
55. Zero oo One
56. Mid Stream Balance
57. Peeling off of Faces of Surfaces
58. Compactified Origins
59. Sathapatya Measuring Rod
60. Inward and Outward Progressions
61. Vedic Mathematics, Science and Technology
62. Vedic Mathematics of Ganita Sutras
63. Artifices Values Chase 'One'
64. Ganita Sutra-1
65. Ganita Upsutra-1
66. Ganita Sutra-2 and Ganita Upsutra-2
67. Ganita Sutra-3 and Ganita Sutra-4
68. Ganita Sutra-5 and Ganita Sutra-6
69. Ganita Sutra-7
70. Ganita Sutra-7 and Ganita Sutra-8
71. Ganita Sutra-9 and Ganita Sutra-10
72. Ganita Sutra-11 and Ganita Sutra-12
73. Ganita Sutra-13 to 16
74. Overview of Vedas reaching us
75. Ganita Upsutras
76. Vedic Mathematics Approach
77. Vedic Mathematics Direction
78. Vedic Mathematics (Ganita Sutras steps for enlightenment)

53

PURE KNOWLEDGE: SELF ORGANIZING VALUE

1. Pure knowledge is known by its basic virtue being the self-organizing value. It is this value which manifests as features of knowledge systems.
2. As the pure knowledge is a boundless domain, as such knowledge system as well are of infinite range. Further as the knowledge system lead to applied values of knowledge, as such knowledge system, on first principles constitute bridges between the pure knowledge and applied knowledge.
3. Pure knowledge being of virtues of affine formats, as such its comprehension even transcends the consciousness states. With it, many times for the sadhkas who have yet to attain transcendental state, have to be dependent upon manifested values comprehension level. This, this way makes them to be more dependent upon the applied values of knowledge or upon the knowledge systems themselves.
4. The ascending order of applied values, knowledge systems and pure knowledge shall accordingly be having ascending order of comprehension levels as to

be of Waking state of consciousness, dream and Deep Sleep states of consciousness and *Turia* (cosmic) state of consciousness of the sadhkas.

5. Knowledge systems as such being of in between position of applied knowledge and pure knowledge, the same parallel to it becomes the in between position of waking state of consciousness and cosmic state of consciousness.
6. This in between stage of waking state of consciousness and consciousness state of consciousness is the sleep state of consciousness, which is of dual formats as being the dream state of consciousness and of deep sleep state of consciousness respectively.
7. The dream state of consciousness is more akin to the applied knowledge domain, while the deep sleep state of consciousness is more akin to the pure knowledge domain. With it, the knowledge systems as well acquire dual states parallel to the formats of dream state of consciousness and deep sleep states of consciousness respectively.
8. The dream state of consciousness format of knowledge system would help transit smoothly to the knowledge system value of domain of applied knowledge. On the other hand the deep sleep state consciousness format would permit smooth transition for the knowledge system to the pure knowledge domain.
9. These features of knowledge systems which on the one hand permit transition to pure knowledge domain and on the other hand to the applied knowledge domain deserve to be chased very gently so that smooth transition may be acquired for its both directional flow.
10. One shall permit one's transiting mind to go step by step starting with the given applied value domain for transcendence there from to the dream state of its

knowledge system. And there from, a step ahead to reach the deep sleep state of the knowledge system. It is only with this attainment, one shall attempt further for the smooth transition from the deep sleep state of knowledge system to the pure knowledge itself of which has been the startwith applied value domain.

11. To have proper comprehension of these transition chase steps, one shall first of all to have evaluation of the values of waking state, dream state, deep sleep state and cosmic state already imbibed by oneself with the help of the ancient wisdom scriptures as well as by having interaction with the senior sadhkas.
12. One may have a pause here and be face to face with one's imbibed values depths of waking state, dream state, deep sleep state and cosmic consciousness states.
13. As these consciousness states are acquirable by sadhana, as such urge should be to acquire these states of consciousness by appropriate steps to distinctively imbibe the values of applied knowledge, two fold states of knowledge systems and of pure knowledge. The consciousness field unfolds itself at core of the intelligence domain of mind.
14. The physical senses domain, intelligence domain and consciousness field are sequential steps in ascending order. As such, these are to be transcended in this sequential order. It is by transcendence through the physical senses domain that the intelligence domain is reached at. And further it is by transcendence through the intelligence domain that the consciousness field stands attained.
15. The sensory domain is of five folds parallel to the five fold, range of senses. It is the interaction range of these five fold senses with the world around that the sensory domain comes into existence. This is known as five steps

long range of five basic elements namely that of Earth, Water, Fire, Air and Space.

16. A step ahead is the Sun. It becomes the source reservoir of virtues and values of intelligence domain as well as of consciousness field. This being so, the Sun, as the source reservoir of intelligence and consciousness becomes the focus of attention for the Sadhkas urging and aspiring for intelligence and consciousness states with whose attainment the knowledge system are to come within the reach of human mind.

◆ ◆ ◆

54
COMPACTIFIED CONSCIOUSNESS STATES

1. Consciousness field evolves at core of the intelligence domain. The consciousness field compactifies its all states. These states of consciousness to begin with are compactified and unified however, despite compactification each of the state is lively for its full expression.
2. Ancient wisdom enlightens us that there is a spectrum of the range of seven states of consciousness, namely, waking state, dream state, deep sleep state, Turia (cosmic) state, transcen-dental state, God state and unity state of consciousness.
3. The phenomenon of compactified range of seven states of consciousness is parallel to the phenomenon of light spectrum visible range of seven colours bands expression as that of rainbow. The consciousness field, as such is parallel to the Sun light field.
4. The consciousness field like sunlight field simultaneously takes care of every band of its spectrum. With it, its system merge as to be of seven parallel systems. Each of these bands and its corresponding system, as

such shall be complementary and supplementary of every other band of the spectrum.

5. The features and values of each of the band, though being an independent and self sustained domain but simultaneously it go exists along with other bands in complementary and supplementary nature and all the bands simultaneously existing being parallel to each other.
6. Further the phenomenon of bands of spectrum being of independents domains and also being complementary and supplementary of each other and further as that those being parallel to each other, also together unify and transit and transform as unified spectrum transcending into the source field values and virtues.
7. It is this characteristics of compactified states of consciousness which deserves to be chased to have comprehension and imbibing of the values of its knowledge systems.
8. Ancient wisdom approaches it as internal expending universe as a field of consciousness and as external expending universe of Sunlight. It is this parallel internal and external features of the expending universe which becomes the expending universe of knowledge as well as of knowledge systems.
9. The light bits, consciousness impulses and intelligence pools in that sequence and order as well as in their reverse orientations shall be coordinating gaps bridging of pure knowledge and applied knowledge in terms of the features of knowledge systems.
10. It is like the affine 'zeros' as pure knowledge and affine 'units' as applied knowledge permitting bridging of gaps in between the two in terms of 'ciphers'.
11. One shall have a pause here and sequentially permit the transcending mind to chase virtues of zeros, values of units and features of ciphers.

12. Ancient wisdom enlightens us that these zeros, ciphers and units mark their independent presence in the solar universe but ahead in creator space (4-space) these indistinguishably super impose upon each other and unified themselves making out the virtues of zeros and values of units to be of the same domain.
13. *NVF (ZERO)=64 = *NVF (UNIT) along the 26 elements range of Vishnu lok/orb of the Sun.
14. One shall have a pause here and permit the transcending mind to face to face with this phenomenon of beginning and end being at the same meant because of transition from linear order to spatial order.
15. It will further help comprehend as that with beginning and end meeting at the same meant would make the middle out and because of it *NVF (CIPHER)=59 = *NVF (SOLID).

Note:- NVF (s) is abbreviation of number value format (s) which are of the range of values 1 to 26 of elements 1 to 26 of Vishnu lok/orb of Sun and parallel to it A to Z respectively get artifices values 1 to 26 in that sequence and order. It is the summation of these values of the letters of the Word like ZERO, UNIT, CIPHER and SOLID that become the NVF (s) of the words. Illustratively the word zero avails four letters namely Z, E, R, O and as such the artifices values, 26, 5, 18 and 15 which together sum up as 26 + 5 + 18 + 15 = 64 and likewise would follow the NVFs of other words.

◆ ◆ ◆

55

ZERO TO ONE

1. Zero to One is a range.
2. Zero and One are the end points of the range Zero to One.
3. Zero, Zero to One and One, as such become three different states.
4. The Zero to One range state shall be having a two fold approach, firstly from the end 'Zero' approaching to the other end, namely 'one', and secondly from the end 'one' approaching to the other end namely 'zero'.
5. The parallel approaches from both ends shall be crossing each other in the mid between at '1/2 value point of the range'.
6. The mid-point, that is 'half value point', may be approached from either end.
7. Both approaches shall be crossing and mid-value point.
8. It may be taken that the approach from the end 'Zero' shall be approaching the mid-value point and shall be trans gracing ahead towards the other end, namely 'one value point'.

9. This would help comprehend as if there is a flow process from mid-value point towards the end value point 'one' as a flow of half range value.
10. Likewise, it would follow that there is a flow process from mid-value point towards the end value point 'Zero' as well.
11. This, this way shall be making out as that there is a two fold flow process of opposite orientations leading towards the pair of end points, namely the end value point 'zero' as well as end value point 'one'.
12. Both these flow streams from the middle value point, though of opposite orientations but at the same time would be of equal values, that is of, half range value each.
13. It would, this way, help tape a two fold flow process from the mid-point and having the complete coverage for the full range in 'half time' then that of the time for the coverage for the range from one end point to the other end point of the range.
14. One may have a pause here and permit the transcending mind to face to face with this mid-point as being the origin.
15. From the mid-point as origin of the range/line/axis, when one shall be transiting to the surface format admitting a pair of axes, it would follow that the flow from the mid/origin point shall be of four folds.
16. Likewise a step ahead within the solid domain of triple axes set-up shall be attaining a six streams flow from the mid of the origin point of 3-space to parallely would emerge a pair of half axis of 1-space, pair of two dimensional frames for 2-space and pair of three dimensional frames of half dimensions for 3-space.
17. It is this working model for approaching the spaces in terms of pair of dimensional frames of half dimensions (axis), which deserve to be chased for their comprehensions as well as for imbibing of their values.

18. This approach shall be enriching the working features of the knowledge systems.
19. A step ahead would emerge the model of approaching spaces domains instead of single centre/origin set-ups to that of a pair of centres/origins.
20. This infact shall be amounting to attainment of transitions from linear order set-ups of space domains to that of spatial order set-ups of space domains.
21. Ancient wisdom even help us attain insight for transiting from spatial order to solid order and even ahead to hyper solid order.
22. With each transition steps, the knowledge systems accordingly would be getting enriched with further features, values and virtues sequentially transcending from of our 3-space world of linear dimensional order to creator space (4-space) of spatial dimensional order and ahead to transcendental domains (5-space) of solid dimensional order and still ahead to self-referral domains (6-space) of creative dimensional order and to unity state (7-space) of transcendental dimensional order.
23. It would be a blissful exercise that one shall sequentially attain transcendence from linear order 3-space to spatial order 4-space solid order 5-space, hyper solid order –4/six-space and finally to hyper solid order 5, unity state domain (7-space).

    It would further be blissful that each stage of this transcendence attainment, one shall have a pause and to have a fresh look about the attained sequential virtues, features and values up to that stage of pure knowledge, knowledge systems and applied knowledge.

◆ ◆ ◆

56

MID STREAM BALANCE

1. The state of origin is that of a mid stream balance.
2. The two fold flow at middle point of zero to one range acquires a balancing state for the middle point as origin.
3. The zero to one range as a flow path of sequential progression from end point zero to end point one, being a west to east positive orientation and as a reverse orien-tation there of from end point one to end point zero from east to west, together as a pair of sequential progressions of opposite orientations shall be creating a balancing state at the middle point for the origin.
4. One shall have a pause here and to have a fresh look at this balancing state of the origin. It shall be, as such creating a zero value state for this state at the middle point for the origin.
5. Simultaneously, it also shall be bringing to focus as that the availability of pair of half ranges values, when stripped off, of their orientations shall be together unifying the absolute value being '1'.

6. This, this way shall be creating simultaneously, a pair of values being 'zero' and 'one' simultaneously for the same state.
7. Still further, it also shall be coming to focus as that as for the whole range from zero to one, there being a pair of sequential progressions of opposite orientations. With their balancing state at the middle it shall be emerging to emergence of zero value progression line in between the pair of +1 and –1 progressions line.
8. It is this spacing at the middle because of zero value line for the +1 and –1 values line lies. The same as such amount to the provisions of surface formats for this set-up.
9. The emergence of four fold values set-up namely as of –1, 0, +1 and 2 shall be making it a four fold manifestation.
10. This four fold manifestation sequentially as such is designated as being the set-up of (–1) value being dimension fold, (0) being boundary fold, (+1) being domain fold and (2) value being the origin fold.
11. It is the simultaneous availability of zero as well as one as values for the mid-point 'origin' which shall be making it a zero value as dimension path for +2 value.
12. One may have a pause here and have a fresh look at the above four fold set-up and comprehend as that line as a representative regular body of 1-space shall be having (–1) value/(–1) space as its dimension fold, zero value/zero space as its boundary fold, +1 value/1-space as domain fold and +2 value/2-space as its origin fold.
13. It would be a blissful exercise to further revisit the above expression from its emergence face till its manifestation as four folds for the zero to one range and to be face to face with this phenomenon of manifestation of line as

manifestation layer format of four folds of values (–1, 0, 1, 2) /(–1-space as dimension, 0-space as boundary, 1-space as domain and 2-space as origin).

14. A step ahead from line to surface, the transition step would be from the manifestation layer (–1, 0, 1, 2) to the manifestation layer (0, 1, 2, 3) .

    Still step ahead it shall be leading to transition for solid into manifestation layers (1, 2, 3, 4).

◆ ◆ ◆

57

PEELINGOFF OF FACES OF SURFACES

1. Surfaces are of pair of faces. The pair of faces of a surface are parallel to each other either of these may the first phase, and the other, in the context shall be the second. To be from one face to the other face of the surface would mean to transcend through even face to the other face.
2. One feature to distinguish the pair of faces of surface would be the orientation as these are of opposite orientation of each other. The transcendence as well from given face to other face, in its reverse process to be back from the other face to the first face, in the context of the first process of transcen-dence may be designated as the process of ascendance.
3. The orientations of the surfaces are to be of circular characteristics. The pair of such orientations may be chased as from north to east and from west to south. If the given face of the surface is of north to east circular orientation then the opposite face of the surface shall be of west to south orientation.

4. This set-up of the faces of the surfaces is the mechanism of the spatial order creator space (4-space) which sandwiches cipher state solids between the pair of faces of the surface. This cipher state linear order solids help peel off the faces of the surfaces.
5. It is this mechanism of creator space (4-space), which deserve to be chased as of four folds, that is, the linear order of solids, spatial surfaces of solids, solid domains and creative origin.
6. These four folds are the four fold set-up of linear order/1-space playing the role of dimension of 3-space, spatial surfaces of solids/2-space in the role of boundary, solid domains/3-space content manifesting solid domain and creative origin/4-space in the role of origin.
7. This four fold manifestation (1-space/dimension, 2-space/boundary, 3-space/domain, 4-space/ origin) as a quadruple (1, 2, 3, 4).
8. The values of artifice 4 being $2 + 2 = 2\times 2 = (-2)\times(-2)$, as such shall be helping us comprehend and imbibe the values of spatial order of creator space (4-space), unifying identical virtues for both faces of surfaces/ spatial order as the positive orientation face shall be leading to $2\times 2 = 4$ and the other face of opposite orientation as well shall be leading to the same values $(-2)\times(-2) = 4$.
9. One shall have a pause here and permit the transcending mind to have a chase of the peeling off mechanism of creator space (4-space) as a process at the dimensional level.
10. There being a spatial dimensional order for creator space (4-space), as such it shall be having a two dimensional frame for its spatial dimension. The pair of opposite orientations of the faces of the surfaces shall be causing a split for the two dimensional frame into a pair of two dimensional frame of half dimensions.

11. These pair of two dimensional frames of half dimensions, as such shall be of opposite orientations parallel to the orientations format of the faces of the surfaces.
12. This split of two dimensional frame into a pair of two dimensional frame of opposite orientations, as such shall be the split of two space domain/surface resulting into the peeling off the pair of faces for their distinct independent roles.
13. One may have a pose here and permit the transcending mind to focus upon the cipher state solid sandwiched in between the pair of faces of the surface. It as such under the split process of the two dimensional domain as well as of its two dimensional frame would remain the sandwiched cipher state solid to be unaffected as, it in the process of the split of two dimensional domain/ two dimensional frame shall be putting it (cipher state solid) as origin of the two dimensional frame.
14. Here it would be relevant to note that 3-space as origin, infact is the origin fold of the four fold manifestation layer (0, 1, 2, 3).
15. The cipher state solid as origin fold being unaffected with the split of two dimensional frame into a pair of two dimensional frames of half dimensions, in fact would mean as that the split of two dimensional frame into a pair of two dimensional frame of half dimension is there because of the cipher state solid origin itself.
16. One shall further have a pause here and have a fresh look at the set-up of the cube as representative regular body of 3-space and glimpse as that the opposite surfaces of cube along its all three dimensions are nothing but the phenomenon of peeling off of the pair of faces of surfaces along all the three dimensions of three dimensional frame.
17. One may further have a pause here and be face to face

with the phenomenon of 0-space being the dimension of dimension of creator space (4-space). As such for split of three dimensional frames itself being in the role of origin of two dimensional frame, and also the origin fold of the manifestation layer (0, 1, 2, 3), the transcendence within creator space (4-space) is firstly is to be to its dimension (2-space) and then to the dimension of dimension (0-space).

It would be blissful to note the features, comprehend the values and to imbibe the virtues of this phenomenon of creator space (4-space) of permitting sequential transcendence and there by permitting unfolding of the creation phenomenon only in steps which is going to be firstly from domain to dimension and then from dimension to dimension of dimension, and it is only thereafter that the step ahead for split of the creative dimensional frame itself may follow but that too because the spatial order of its dimensional order, it shall be along its first dimension, as well as its second dimension shall be permitting independent transcendence flow whose carriers shall be along independent paths and thereby there shall be elusive state for the mind unless and untill the mind is in a unifying state capable of glimpsing the unifying phenomenon of the transcendence stream which are also of self-referral features in terms of which the self-referral (6-space) is attainable and not otherwise.

◆ ◆ ◆

58

COMPACTIFIED ORIGINS

1. Peeling off of orientations of line and consequential emerging of four fold manifestation format for line/ interval/1-space as of quadruple values (–1, 0, 1, 2) makes out value '2'/surface/2-space as the origin/ origin fold.
2. The peeling off the pair of faces of the surface leads to the four fold manifestation layer of quadruple values (0, 1, 2, 3) with value '3'/3-space as origin/origin fold.
3. Let us have a fresh look at the set-up of the cube (representative regular body of 3-space). We shall be observing as that it has linear axes/1-space in the role of its dimension. Further as that surfaces/2-space playing the role of boundary fold. The solid content/volumme, as such is 3-space itself playing the role of a domain fold. Still further as that the permissible cut of 3-space as eight octants and parallel to it the split of cube into eight sub cubes is there because of centre of the cube being the seat of 4-space as origin/origin fold.
4. This as such lead us to a four fold manifestation layer

with 1-space as dimension fold, 2-space as boundary fold, 3-space as domain fold and 4-space as origin fold, and parallel to it there being a quadruple values (1, 2, 3, 4).

5. This sequential set of quadruples, namely (–1, 0, 1, 2), (0, 1, 2, 3) and (1, 2, 3, 4) leads to the existence phenomenon of compactified origins/origin folds. 2-space in the role of origin of 1-space, 3-space in the role of origin of 2-space and 4-space in the role of origin of 3-space, leads to formulation as that (n + 1) space being in the role of n space for all values of n.
6. This formulation of n + 1-space being the origin of n space, as such becomes the formulation of compactified origins with the feature of inward sequential unfolding expressions for the manifested creations.
7. The manifested creations being of four folds, namely, dimension fold, boundary fold, domain fold and origin fold, and that too as sequential order of consecutive set of four dimensional spaces with general formulation (n – 2, n – 1, n, n + 1) lead to hyper cubes format for the manifested creations.
8. Interval as representative regular body of 1-space is of hyper cube 1 format with (–1) space content as dimension fold, 0-space content as boundary fold, +1-space content as domain fold and 2-space content as origin fold.
9. A step ahead, the square as a representative regular body of 2-space is of the format of hyper cube 2 with 0-space content manifesting as its dimension fold, 1-space content manifesting as its boundary fold, 2-space content manifesting as its domain fold and 3-space content manifesting as its origin fold.
10. It would be a blissful exercise to chase cube, the representative regular body of 3-space, being as of 1-space content manifesting its dimension fold, 2-space

content manifesting as its boundary fold, 3-space content manifesting as domain fold and 4-space content manifesting as its origin fold.

11. The value n = 4, shall work out four fold manifestation layer being a quadruple values (2, 3, 4, 5) as the set-up of 2-space content as dimension fold, 3-space content as boundary fold, 4-space content as domain fold and 5-space content as origin fold.
12. Let us have a fresh look at the set-up of interval, square and cube and comprehend as that there are two boundary component of zero space content for the interval, four 1-space content boundary components of square and six 2-space content boundary components of cube. A step ahead would be eight 3-space content boundary components of hyper cube 4.
13. There being eight 3-space content (cubes) as boundary components of hyper cube 4 would help settle the formation of the hyper cube 4. The split of 3-space into eight octants and parallel to it the split of cube into eight sub cubes would help comprehend the centre of cube being of hyper cube 4 format as that the centre of the cube is enveloped by eight sub cubes of cube. It would be a phenomenon parallel to the origin of 3-space being enveloped by eight octants of 3-space.
14. A step ahead it can be well projected and comprehended as that hyper cube 5 shall be accepting ten 4-space content boundary components, and as such the boundary emerges to be the creative boundary because of 4-space content manifesting here as boundary.
15. As a sequential step, hyper cube 6 accepts 12 boundary components of 5-space content, and as such boundary of hyper cube 6 becomes the transcendental boundary as the content of 5-space is of transcendental features.

16. Parallel to linear axis emerge manifested axis. It would be of the manifestation layers formats. It is of the values of hyper cubes. It shall be amounting to linear axis frame of cube transiting and transforming into manifested linear dimensions frame for hyper cube 3.
17. The manifested linear dimension shall be of the format of four folds of interval formats (0 to 1), (1 to 2), (2 to 3).
18. This as such shall be working out a manifestation layer (0, 1, 2, 3).
19. The transition from linear order to spatial order, shall be requiring as many as four dimensions (instead of three dimensions of 3-space).
20. These four dimensions shall be four sequential quadruples (0, 1, 2, 3), (1, 2, 3, 4), (2, 3, 4, 5) and (3, 4, 5, 6).
21. These four manifestation layers together shall be availing 4 4 matrix/grid format for the manifestation layers format of spatial dimensional order being of arrangement and set-up as follows:

0	1	2	3
1	2	3	4
2	3	4	5
3	4	5	6

22. This arrangement and set-up shall be constituting 9 grid zones on the grid zone format which itself shall be the tenth source reservoir grid zone, to be designated as the origin source reservoir grid zone, in short to be the source origin or simply the ORIGIN.
23. Simultaneously, it would be relevant to note that this arrangement and set-up of spatial grid format for the spatial order of spatial manifested dimensions is availing in all seven spaces range (0, 1, 2, 3, 4, 5, 6).
24. As seven points may count six units (length units), it

shall be the coordination of 1-space to 6-space.

25. This spatial dimensional order of creator space (4-space) deserves to be chased as capable of covering Earth to Sun (Earth, Water, Fire, Air, Space, Sun) range of solar universe.

    The line zones values of this arrangement and set-up of spatial dimensional order shall be leading to the ultimate Brahman Vidya as well as Par Braham enlightenment.

◆ ◆ ◆

59

SATHAPATYA MEASURING ROD

1. *Athrav Ved* is of applied values of Vedic knowledge its upved, namely the *Sathapatya Ved* is the scripture of knowledge systems. The measuring rod for chase of the knowledge systems along their formats is designated and is known as *Sathapatya* measuring rod.
2. The sathapatya measuring rod is the synthesis of hyper cubes 1 to 6 being the manifested values of the representative regular bodies of 1-space to 6-space.
3. It is in terms of this *sathapatya* measuring rod that the whole range of existence phenomenon from Earth to Sun is chased by the Vedic systems.
4. Both, the inward as well as outward expressions of the existence phenomenon chased in terms of the sathapatya measuring rod.
5. Even the Pursha format along which manifests human frame as well as the human body is also chased by Vedic systems in terms of the Sathapatya measuring rod.
6. The existence phenomenon along its ten directions as well is chased direction wise, as well as collectively, in terms of Sathapatya measuring rod.

7. The sathapatya measuring rod being synthesis of hyper cubes 1 to 6 (interval, square, cube, hyper cube 4, hyper cube 5 and hyper cube 6) as representative regular bodies of 1-space to 6-space as sets of their four fold manifestation values, are also chased collectively, as well as fold wise, in terms of the sathapatya measuring rod.
8. As dimensions fold, –1-space, 0-space, 1-space, 2-space, 3-space and 4-space shall be playing the roles of dimension fold of respective dimensional domains of 1-space to 6-space respectively and so is going to be the 6 steps long chase of the measuring rod.
9. As the 0-space, 1-space, 2-space, 3-space, 4-space, 5-space shall be playing the roles of boundary folds of interval to hyper cube 6, as such, parallel to it, would be the six steps long chase of boundary folds in terms of the *sathapatya* measuring rod.
10. Further as 1 to 6-space shall be the domain folds so their chase in terms of *sathapatya* measuring rod.
11. Still further as 2-space to 7-space shall be playing the roles of origin folds, as such parallel to it six steps long would be the chase thereof in terms of sathapatya measuring rod.
12. In general manifestation are of four folds, namely dimension fold, boundary fold, domain fold and origin fold, while the measuring rod is of six steps long range, as such in first three steps of the measuring rod, dealing namely, interval, square and cube together would permit simultaneous chase as a single step in terms of cube itself as interval, square and cube together can find expression within cube itself.
13. This expression and chase of interval, square and cube within the cube is the unique feature of the cube within creator space (4-space).

14. This feature of the cube giving rise to the simultaneous existence for interval, square and cube is designated and is known as the creative feature of the cube as a domain fold.
15. In fact this creative feature of the domain fold is not only restricted to the cube within creator space (4-space) but is also marking its presence in reference to every other hyper cube within creator space (4-space) as well.
16. This creative feature of the domain folds of hyper cube is there because of the spatial dimensional order of the creator space (4-space) itself.
17. One may have a pause here and permit the transcending mind to have a fresh look at the set-up of cube as hyper cube 3 of linear dimensional order and chase such structural set-up being permissible along each of the two linear dimensions of the spatial order of the creator space (4-space).
18. As such it may be comprehended and imbibed the values as that the dimension fold, the boundary fold and domainfold are simultaneously available within the domain fold itself.
19. With this feature of dimension fold, domain fold and boundary fold being simultaneously available within the domain fold itself, as such it makes the *sathapatya* measuring rod being the creative measuring rod of four folds parallel to third to sixth step of *sathapatya* measuring rod.
20. Here it would be relevant to note that this creative feature of simultaneously creating dimension fold, boundary fold and domain fold within domain fold itself, is parallel to the creation due to the first axis, creation due to the first pair of axis and creation due to all the three dimensions of 3-space.
21. One shall further have a pause and to permit the transcending mind to remain in deep trans to glimpse

and to be face to face with this phenomenon of Earth as physical phenomenon of 3-space set-up of linear dimensional order.

22. With 3-space as dimension, 4-space as boundary fold and 5-space as domain fold would further under the same creative order shall be folding these three folds within 5-space as a domain fold with 6-space as origin.
23. The four fold parallel for manifestation paths of sequential progressions of dimensions fold, boundary fold, domain folds and origin folds may be chased individually or in pairs or in triples or all together as quadruples as unified flow path in terms of the *sathapatya* measuring rod.
24. The transition and shift from linear dimensional order to spatial dimensional order, as such would be parallel to transition from 1 as 1 to 2 as 1. With this transition the counting range (1, 2, 3, 4, 5, 6) and so on shall be transiting and transforming as the counting range (2, 4, 6, 8, 10, 12) and so on. This transition would be taking from dimension fold to boundary fold.
25. The counting range (2, 4, 6, 8, 10, 12) is permitting a jump at every step for the counting range (1, 2, 3, 4, 5, 6, 7, 8, 9, 10, 11, 12). It is like a shift to the even numbers, and as such there being a jump over odds and as such the counting range would split into a pair of counting ranges of odds and evens.
26. This split of counting range into a pair of counting ranges of odds and evens, infact are the sequential ranges because of sequential shifts from dimensions fold to domains fold like 1-space as dimension leading to 3-space as domain, 3-space as dimension to 5-space as domain and so on. Likewise there would be a shifts from 2-space as dimension to 4-space as domain, 4-space as dimension to 6-space as domain and so on.

27. Along these formats, the range of five basic elements together with Sun shall be providing a pair of approach paths, firstly, from Earth to Fire to Space and secondly from water to air to Sun and chase like that in terms of *sathapatya* measuring rod works out the different features of our existence phenomenon and corresponding ranges of knowledge systems.

◆ ◆ ◆

60
INWARD AND OUTWARD PROGRESSIONS

1. The features of simultaneous inward and outward progressions of the manifestation format of our existence phenomenon deserves to be chased as features of the knowledge system of creator space (4-space).
2. Within creator space (4-space), whole range of creations are of four fold manifestation format of idol of the Creator the supreme, the four head lord, Lord Brahma.
3. This manifestation format makes creations as four fold manifestation layers of synthesis of dimensional contents of four consecutive dimensional spaces.
4. Illustratively, cube, representative regular body of 3-space as hyper cube 3 being the manifested creations within 4-space is of four folds namely (1-space as dimension, 2-space as boundary, 3-space as domain and 4-space as origin).
5. 4-space itself being of a spatial dimensional order as such hyper cube 4 itself expresses as a manifestation

layer (2-space as dimension, 3-space as boundary, 4-space as domain and 5-space as origin).

6. One may have a pause here and have a fresh focused attention upon the transition and transformation of features from that of cube as hyper cube 3 to hyper cube 4.
7. Here one may comprehend and imbibe the values of this transition and transformation steps as taking from 1-space as dimension/linear order to that of 2-space as dimension/spatial order.
8. As this transition and transformation process for cube/ linear order is culminating at its centre/origin fold being of spatial order format, as such it would amount to inward progression process.
9. Now, on the other hand the manifestation layer set-up of the cube as of 1-space as dimension and 2-space as boundary fold will bring to focus as that in the process of transition from dimension fold to boundary fold.

   There is progression from the role of 1-space to the role of 2-space which is of external progression feature as that this is culminating at the boundary, that is, at the external frame of the domain fold.
10. This progression process for 1-space (as dimension) and 2-space (as boundary), is designated as the external progression. The internal progression is taking place at the centre/origin of 3-space while the external progression is taking place at the frame/boundary.
11. One may have a pause here and note that the external progression is coordination of dimension fold with boundary fold while the internal progression is coordination of dimension fold with origin fold.
12. These values deserves to be comprehended thoroughly and to be imbibed fully.

13. The chase of internal and external progression along the *sathapatya* measuring rod are to be the conscious steps as in each case special features are to be taken into account.
14. The external progression chase as such is to be of the steps (*i*). one space as dimension to 2-space as boundary (*ii*). 2-space as dimension to 3-space as boundary and so on.
15. The internal progression chase is to be of the steps (*i*). 1-space as dimension to 2-space as dimension of origin (*ii*). 2-space as dimension to 3-space as dimension of origin and so on.
16. The external progression shall be as going from boundary to boundary of the hyper cubes but at the same time it shall be bringing to focus as that the gaps in between the pair of progression steps shall be the dimensional domains.
17. On the other hand the internal progression of the orders and values from linear order to spatial order and so on would be attainable in terms of their corresponding dimensional domain which shall be interlinked as domains and origins.
18. The phenomenon of internal progression as of the formats and steps leading from domain to its origin is designated and known as the transcendence phenomenon.
19. In its reverse orientation, the steps of reaching from origin to domain is designated and known as ascendance phenomenon.
20. As the origin is of next higher dimensional order then that of the domain itself, as such the higher order content flow would be there from the origin into the domain, which would mean that the lower order content gets super imposed by the higher order content.

21. One may have a pause here and have a fresh look at the manifested values of the set-up of hyper cubes and comprehend its values as that boundary is of lower content order then that of the domain content order which would mean that higher content order can be wrapped within the lower content order.
22. Now, the ascendance shall be imposing higher content order upon the content order.
23. The transcendence from the boundary to domain and ascendance from the origin to the domain being permissible shall be making the existence phenomenon of the domain as to be of parallel but of opposite orientation flow paths.

    This may help us appreciate how the manifested creations because of its parallel but of opposite orientations flow paths is sustaining life and death within human frame and all other frames.

◆ ◆ ◆

61

VEDIC MATHEMATICS, SCIENCE AND TECHNOLOGY

1. Vedic mathematics, Science and technology is of the format of range of Earth to Sun.
2. The Earth to Sun range is of six steps coverage being EARTH, WATER, FIRE, AIR, SPACE and SUN.
3. Earth to Sun six steps are the manifested domains of linear, spatial, solid orders known as *'TRILOKI'* and hyper orders 4, 5 and 6 known as *TRIMURTI.*
4. *Triloki and Trimurti*, within creator space (4-space) are linear order manifestations while the *Trimurti* are spatial order manifestations. A step ahead within transcendental space (5-space), which itself is of solid order, the *Triloki* and *Trimurti*, accordingly go transcendental. And, a step ahead within self-referral space, everything automatically goes self-referral.
5. Sadhkas fulfilled with the intensity of urge to glimpse features of *Triloki* and *Trimurti* as within self-referral space (6-space) shall sequentially approach *Triloki* and *Trimurti* as such as in terms of linear order, spatial order, solid order and hyper solid order.

6. The sole enlightened beings fulfilled with intensity of urge for *Braham Vidya* shall chase *Trimurti* in unity state of transcendental order.
7. Beyond that is the Par Braham of self-referral order and all that of what it unfolds itself.
8. As such, the sadhkas fulfilled with such intensity of urge shall initiate themselves on the unison paths of *Sankhiya nishtha* and *yoga nishtha.*
9. For it, one is to go simultaneously availing artifices of numbers presuming the existence of geometric formats on the one hand and also availing dimensional frames presuming the existence of artifices of numbers.
10. This unison path, initiation may be had in terms of the values and virtues of Ganita Sutras (including *Ganita Upsutras*) but for it, it would be better if one comprehends well the features of Vedic alphabet format having reached us in Devnagri script.
11. The *vedic* alphabet format features, particularly the placement about the placement of the individual letters in the organization set-up of the Vedic alphabet format.
12. It is the placement features of the individual letters together with the script form of the letters which together would help have insight about the frequencies (of sound with potentialities of their transition into those of light) would deserve to be comprehended thoroughly and also to be imbibed fully to reach at the values and virtues of the Ganita Sutra (mathematics hymns).
13. One shall permit the transcending mind to remain deep in trans to glimpse and to be face to face with the values and virtues of the Ganita Sutras.
14. It is this insight which shall be making one conscious of the basics on first principles being workable in the

sequential order of the individual letters being availed by the text of every Sutra.

15. This way, the beginning with the very first letter of the very first Ganita Sutra namely the sixth vowel, one shall sequentially having coverage of the text of Ganita Sutra–1 in sixteen steps. Like that the chase path shall be in the sequential order of the Ganita Sutras.
16. Parallel to it would follow the complete chase for Ganita Upsutras as well, as an independent scripture and also as being the unified part of the complete text of Ganita Sutras and Upsutras being the single scripture. As a single unified scripture covering the text of Sutras and Upsutras exhaustively would avail sole syllable Om (ॐ) as the source reservoir and *Parnava* प्रणवः as the fruit of this transcendental chase.
17. The transcendental chase of Ganita Sutras (including Upsutras), as such is of 520 letters range with sole syllable Om (ॐ) being its first letter and *Parnava* प्रणवः as transcendental fruit of the values and virtues of this transcendental range, being the last letter.
18. This 520 letters chase is to provide us the transcendental path (of solid order) for beyond knowledge, knowledge system as well as for the applied values, as a discipline of Vedic mathematics, science and technology being the single discipline for whole range of needed values for our existence phenomenon within the solar system.

◆ ◆ ◆

62

VEDIC MATHEMATICS OF GANITA SUTRAS

1. Ganita Sutras is a complete scripture in itself.
2. Ganita Sutras include Ganita Uputras.
3. Being independent scripture of Vedic knowledge, it accepts sole syllable Om (ॐ) as its source reservoir of values and virtues.
4. Further as that its end fruit is the *Parnava* प्रणवः.
5. The beginning to end range of values and virtues flow path through the artifices of Sutras (hymns), as such is of 31 steps (Om (ॐ) + Sixteen Sutras + thirteen Upsutras + end fruit value *Parnava* प्रणवः).
6. This entire range is of 520 letters in all.
7. This, this way is of 520 values range.
8. The Vedic knowledge and its values as the Discipline of Vedic mathematics, science and technology, avail format of features whose basics on first principles are of the range of constituent virtues of the values of Ganita Sutras.

9. These constituents virtues of the values of Ganita Sutras, in modern language may be approached as self evident truths being accepted as the axioms and postulates of logical steps for structural layout of knowledge systems to bridge the gaps between pure knowledge and applied knowledge domains.
10. Taking that way sixteen Sutras would emerge as sixteen axioms range and thirteen Upsutras as thirteen sub axioms while 520 letters availed by the text to cover from beginning to end, as such would emerge working postulates.
11. Further as that, the Ganita Sutras text being the composition, it shall be having its organization format as well as text being a manifestation thereupon. Accordingly to approach it the features of the organization format as well as the values of the manifestation there upon of the text are to be separately taken care of.
12. As such, each Sutra, in itself would emerge as a complete scripture. Thereby the organization format as well as manifestation there upon of each Sutra shall be of distinct sets of features of the organization format as well as there would be a distinct sets of values of the manifested text of the sutra. The virtue of these features and values would express as the self-evident truth as the working rule of the Sutra.
13. Each sutra, as well as each Upsutra as an axioms/sub axioms would deserve to be comprehend fully to be imbibe the virtues of the features and values of the organization format as well as of the text manifesting thereupon giving rise to the working rule of the Sutra.
14. Ganita Sutras, being sixteen in number shall deserve to be chased for their comprehension and for imbibing of values in the sequence and order of the Sutras but each

Sutra as such is to be approached as a self contained complete scripture.

15. Likewise would deserve to be approached the Ganita Upsutras as well. A pair of sixteen axioms and thirteen sub axioms ranges parallel to sixteen Sutras and thirteen Upsutras shall be requiring their chase in 29 steps.
16. However, throughout this chase, and at each steps of its chase one is to ever remain conscious as that the chase steps are in reference to the start with source reservoir of sole syllable Om (ॐ) and fruit of the chase being the *Parnava* प्रणव: formulation.
17. As there would be a very wide spectrum of virtues, values and features being associated with every Sutra and Sub sutra, as such the chase can be aspect wise like that of the Discipline of artifices of numbers and of dimensional frames.
18. This aspect wise split of the spectrum is like that of colours spectrum of the light and each aspect like distinct colour of the spectrum shall be working out a distinct frequencies domains taking aspect wise chase as same frequency chase from Sutra 1 to Sutra 16 and from Upsutra 1 to Upsutra 13 shall be a blissful experience.
19. To begin with, one may have chase for the frequencies values of artifices of numbers and then there after the chase may be about the frequencies values of the dimensional spaces.

    A step ahead it would be very blissful to have simultaneous chase as a unified field of the frequencies values for artifices of numbers as well as for the dimensional spaces.

◆ ◆ ◆

63

ARTIFICES VALUES CHASE 'One'

1. One shall sit comfortably and permit the transcending mind to comprehend and imbibe the values of 'one'.
2. 'One', sole, loan, alone, single, whole, only one manifest some of the features of 'one'.
3. One shall sit comfortably and permit the transcending mind to chase the expression for one being '1', as of the format of a vertical line/vertical interval, a flow path.
4. One shall further permit the transcending mind to chase the expression for one being '1', '01', '001'; '1^0', '2^0', '3^0'; '1^1', '1^2', 1^3'; unit, unit length, unit area, unit volumme; $1/1$, $1 \times 1\ /1$, $1 \times 1/1 \times 1$; $1 - 0$, $2 - 1$, $3 - 2$, ; $1 \times 1 - 1$, 2 $1 - 2^0$, and the like features of '1'.
5. One shall still further permit the transcending mind to chase 'one as whole', and 'part as well one'.
6. Still further one shall permit the transcending mind to chase 'part being equal to 'whole'.
7. Still further one shall permit the transcending mind to chase 'one as one', 'two as one', 'three as one' and so on. It would be a blissful exercise to chase '1 as 2' and

'2 as 1', individually as well as simultaneously.

8. Pair as one pair, angle as one angle, triangle as one triangle, quadrilateral as one quadrilateral, pentagon as one pentagon and so on shall be giving us insight about the internal self expending features of the set-up of '1'.
9. It is this internal self-expending expression for the set-up of one deserves to be chased as of one ultimately unfolding as infinity.
10. One shall sit comfortably and permit the transcending mind to chase the emergence of 'infinity' in finite.
11. One shall chase line at infinity transiting and transforming into circumference of a circle, and circumference of a circle transiting and transforming into surface of a sphere, and so on sphere as boundary of hyper sphere.
12. Transition from a line to a pair of lines, being parallel to each other, shall be leading to a pair of flow paths.
13. It would be a blissful exercise to simultaneously chase '1' and '11', as a pair of parallel flow lines of artifices of numbers.
14. One may have a pause here and permit the transcending mind to completely comprehend and to fully imbibe the values and virtues of the equality '=' symbol, a pair of parallel lines of equal lengths.
15. One may further have a pause and to permit the transcending mind to chase line as a track of a moving point yielding line as a set-up of points; and 'one' as a set-up of 'zeros'.
16. One may further have a pause and permit the transcending mind to chase the symbol of division '/'.
17. One shall be comprehending here as that the two point object, being a single pair of points/objects, are split/

separated by the intervening flow line, which may be designated as a separation line.

18. One shall have a pause here and to have a fresh look at the conceptual format of the symbol of division splitting a pair into singles in terms of separation line.
19. It would be like of the two (pair) as a pair of arms of an angle, splitting both arms and making them singles which previously being synthesized as a pair by the single point (vertices of the angle).
20. This separation line as such becomes the symbol of '-'.
21. One may further have a pause and permit the transcending mind to comprehend the interlinking values of division and subtraction, as much as to divide actually means to have repeated subtractions. Further with the help of second degree value being divided by first degree value leading to a single degree value as the degree of the divisor gets subtracted from the degree of the dividend, would give us an insight for the separating line transiting and transforming as an expression symbol for '-'.
22. One may further have a pause here and permit the transcending mind to comprehend the interlinking features of subtractions and addition operations.
23. One shall have a subtraction look at the symbols of '–' and of addition '+'. One may note that while the minus symbol is availing single line, the addition operation avails a pair of lines.
24. It is like reaching from one to two '2'.
25. One shall again have a fresh look at the expression of symbol of addition '+', being the synthesis of horizontal and vertical lines.
26. It is a set-up of a pair of axes/ dimensional frame of plane/2-space.

27. It as such shall be amounting to transition and transformation from the format of a line to the format of a plane/ from 1-space format to that of 2-space format.

28. Here one shall again have a pause and to permit the transcending mind to comprehend the interlinking features of addition and multiplication operations.

29. Multiplication as repeated addition is one feature of interlinking of addition and multiplication operations. The addition of degrees of the pair of values under multiplication operation is another interlinking feature of addition and multiplication operations.

30. In this background, when one shall be having a fresh look at the symbolic expression for addition '+' as well as for multiplication ' ', one shall be noticing that it is the shift for the pair of axes avail for addition operation as being parallel to the sides of the square into that of the diagonals of the square.

31. This shift from sides of the square into diagonals of the squares deserves to be comprehended fully and these values to be imbibed completely.

32. It is the comprehension of the shift implications for the two dimensional frame/pair of axes shall be perfecting insight about the interlinking of the addition and multiplication operations for artifices of numbers being parallel to the corresponding dimensional frames.

33. It also would as such, give us insight from the four directions frame of north, east, south, west and transition and transformation there from to that of eight directional frame (a set-up of four directions and of four sub directions, namely north – east, east – south, south – west and west – north).

34. It would be a blissful exercise to chase this shift and consequential transition and transformation for the four directional frame into eight directional frame as structuring and deciphering out of pair of faces of the plan.

35. With it follow the emergence of volumme in between the pair of faces which is to manifest as a cube.

36. It is the creation process.

37. It is the process of transition and transformation from the linear order to the spatial order.

38. This transition from linear order to spatial order is also to provide a shift from linear order 3-space domain to spatial order 4-space domain. It is an affine state.

39. It is full in itself.

40. It would be a full volumme.

41. It is sole syllable Braham lively as a volumme (of cube) enveloped within a synthetic set-up of eight corners, 12 edges and 6 surfaces.

42. One shall sit comfortably and permit the transcending mind to be face to face with the values and virtues of ancient wisdom enlightenment : 'ॐ इति एक अक्षर ब्रह्मम' *Om Itah Ek Akshar Braham/Om* (ॐ) is of end values and virtues 'sole syllable Braham'.

43. It further would be blissful to chase 26 elements alphabet of artifices 1 to 26, admitting pairing operation yielding number values formats for A to Z as 1 to 26 in that sequence and order and in terms of the pairing operation of the values of the number value formats of individual letters i.e. one for a as its number value format, two for be as its number value format and so on shall be leading to number value formats, in short NVFs for words formulations.

44. It would be blissful to chase NVF (SOLE)=51 = NVF (FULL), NVF (SYLLABLE) = 88 = NVF (VOLUMME) and NVF (BRAHAM) = 41 = NVF (AFFINE).
45. NVF (OM) = 28 = NVF (AIR); Artifice 28 is the second perfect number. Air is the fourth element. Its hyper cube 4 format shall be of spatial dimensional order that is two space in the role of dimension of hyper cube 4, the representative regular body of 4-space/creator space (4-space).
46. One shall permit the transcending mind to be face to face with the enlightenment of ancient wisdom : 'ॐ इति एक अक्षर ब्रह्मम' *Om Itah Ek Akshar Braham'*; and enlightenment of artifice 'one'.

◆ ◆ ◆

64
GANITA SUTRA-1

1. Text

SUTRA - I

एकाधिकेन पूर्वेण

01	02	03	04	05	06	07	08	09	10
ए	क्	आ	ध्	इ	क्	ए	न्	अ	प्
11	12	13	14	15	16				
ऊ	र्	व्	ए	ण्	अ				

Total Letters	Vowels	Nasels	Consonents
16	8		8

2. Format Outline of Working Rule

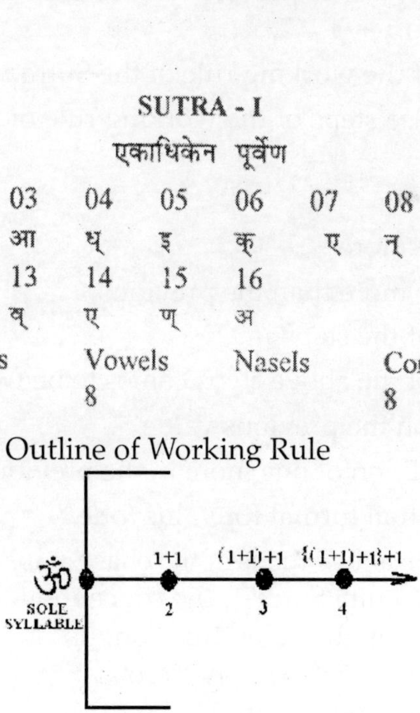

3. Initial Steps Outline
   (*i*) Read the text of the Sutra.
   (*ii*) Pronounce the text Loudly.
   1. Sequentially tabulate the letters of the text.
   2. Go through the working rule of the Sutra.
   3. Translate this working rule in the language of numbers, as:

      For 'n' as previous number, reach at 'n + 1'.
   4. Illustrations

      For n = 0, as previous number, rule shall be taking us to 0 + 1 = 1.

      For n = 1, as previous number, rule shall be taking us to 1 + 1 = 2.
4. Steps of the working rule of the Sutra.

   The chase steps of the working rule of the Sutra come to be:

   (*i*) One

   (*ii*) One more

   (*iii*) One more than the previous
5. Focus of the steps

   Focus of the above steps comes to be two fold

   (*a*) Upon the previous value

   (*b*) Addition of one more in the previous value
6. Conceptual format for value 'one'.

   From the above focus of the chase steps of the working rule of Ganita Sutra-1, the conceptual format of value 'one' comes to be as that 'one is one more than its previous value', namely 'zero'.
7. One as a starting point.

   In case, 'one' is taken as the starting point, then its value which comes to focus under the rule of Ganita Sutra-1, is that it is the value, which is more than its previous state value.

8. One as a unit

   This value unit is at takes us from 'zero' to 'one'.

9. Negative Unit one

   Let us have a pause here and permit the transcending mind to glimpse the features of the emergence of the unit value taking us from zero to one as the positive unit value. Now when 'one' attempts to comprehend the value of 'zero' in terms of the positive unit value, one shall be coming to face to face with the previous state for zero as being such as that from their positive unit shall be taking us to 'zero' itself. It, as such shall be making it to be the previous state for zero as being of negative unit value as from their in terms of positive value, one shall be attaining 'zero'.

10. Zero sandwiched between negative unit and positive unit values.

11. One shall sit comfortably and permit the transcending mind to comprehend the phenomenon of 'zero' being sandwiched between negative unit value and positive unit value.

12. Continuity for the numbers line.

13. One may further have a pause and permit the transcending mind to remain in deep trans to glimpse the continuity of the numbers line of positive units as well as the negative units, simultaneously being sequentially arranged in terms of the working rule of Ganita Sutra-1 as to be of the steps of 'one more than the previous 'one'.

14. Zero as a whole number.

15. One may further have a pause and permit the transcending mind to glimpse the features of continuity of numbers line and to be face to face with 'zero' as

being the whole number like the positive and negative units values counts.

16. Positive and negative orientations

    The set-up of the numbers line with the focus upon 'zero' as having positive units on its one side and negative units on its other side shall be indicating opposite orientations sequences for the positive unit values and negative unit values.

17. Neutralization of orientations at 'zero'

    One may have a pause here and permit the transcending mind to have a fresh look at the set-up of the numbers line and to comprehend the emergence of unification of orientations amounting to neutralization of orientations at 'zero'.

18. 'Zero' as a fluctuating point of numbers line.

    As the working rule of Ganita Sutra-1 as to be of steps 'one more than the previous', is to be of infinite units, as such 'zero' is to be of a fluctuating point seat on the numbers line, as from any point of the line there can be infinite points from the given reference point of the line.

19. Stripping off of orientation of a line

    With 'zero' value along the numbers line being a fluctuating point, and it can be as far of towards west along the west to east line, and as the whole range of numbers along the west to east line from the given reference point would be the positive units, as such it shall be making the whole line from the far off west till east as being the positive units range and hence as to be of positive orientation format of a line. Like wise the fluctuating seat of 'zero' when would be towards the far off east, it shall be making the whole range from far of east till the west as to be the range of negative range values and hence as to be of negative orientation format. And with it the positive units values range and the negative values range shall be stripping off positive and negative orientations of a line.

20. Ganita Sutra-1 and 14

    One may have a pause here and permit the transcending mind to comprehend the features of organization format of Ganita Sutras which put the Ganita Sutra-1 as to be of steps 'one more than the previous one' and Ganita Sutra-14 of the working rule of steps 'one less than the previous one' as being distinct sutras and being separated in the organization format as being of placements of first Sutra and fourteen Sutra with intervening gap of 12 Sutras namely Sutras 2 to 13.

21. Ganita Sutra-1 as the source Sutra.

    One way to approach Ganita Sutra-1 is as the source Sutra as that sequential transitions from Sutra-1 itself to Sutra-2 and on ward up till Sutra-16 is to follow from the internal expansion parallel to the sequential order of letter 'one' to letter 'sixteen' of the text of Ganita Sutra-1 itself.

22. Artifices of numbers

    The other way to approach Ganita Sutras is in terms of the artifices of numbers for whose organization the basic rule is supplied by Ganita Sutra-1 itself as that the exhaustive coverage for number systems is ultimately to be worked out by the steps of 'one more than before'.

23. Geometric formats approach

    The other approach to the Ganita Sutra is as of geometric formats of dimensional orders in terms of which the sequential organization of dimensional spaces can be exhaustively chase for the coverage of entire universal existence phenomenon.

24. 'Zero' unit as dimensional value

    One may have a pause and permit the transcending

mind to comprehend completely and to imbibe fully the values of 'zero' as of unified seat of neutralization of pair of orientations. It would be like the pair of axes neutralizing and exhausting themselves as to be of zero unit value, or looking the other it would amount that the zero unit value shall be emanating a pair of orientations from the common source and these pair of infinite flow paths shall be manifesting a surface/2-space set-up within a pair of dimensional set-ups of 'zeros (values)'.

25. Continuity of dimensional space

    One shall further have a pause and permit the transcending mind to glimpse the phenomenon of emergence of positive and negative dimensional spaces which along with zero unit value 'two' dimensional space to constitute a dimensional continuity of the feature that the difference of the values of dimensional units and that of domain units to be two parallel to the difference between the value of surface/two space and its dimensional order being of zero unit value.

26. '–1' space as dimensional order of '+1' space.

    One shall have a pause and have a fresh look at the set-up of line/interval as representative regular body of one space. While one shall be permitting one's transcending mind to remain in deep sittings of trans one shall be glimpsing and shall be face to face with the phenomenon of striping off of positive and negative orientations as the formats for positive units and negative units ranges. It shall be helping comprehend and to have insight as to the difference between '+1' and '–1' as to be of two units value because of which '–1' space shall be playing the role of dimension of '+1' space.

27. Phenomenon of parallel artifices of numbers and of dimensional frames.

28. One shall have a fresh look at the working rule of Ganita Sutra-1 and to permit the transcending mind to have an insight of the ancient wisdom of unified values of *Sankhiya Nishtha* and *Yoga Nishtha* as that the artifices of numbers presume the existence of dimensional frames and on the other hand *Yoga Nishtha* the existence of dimensional frames presume the existence of artifices of numbers and both getting organization values to be regulated by the rule of Ganita Sutra-1 as to be the steps 'one more than the previous one'.

◆ ◆ ◆

65

GANITA UPSUTRA-1

1. Text

आनुरूप्ये Anurupyena
Proportionately

2. Initial Steps Outline:

(*i*) Read the text of the Sutra

(*ii*) Pronounce the text Loudly.

(*iii*) Sequentially tabulate the letters of the text.

(*iv*) Go through the working rule of the Upsutra.

(*v*) Translate this working rule in the language of numbers, as proportionally.

With formulation n being replaced by m

(*vi*) Illustrations

Replace '1' by '2'. It shall be making the rule 'one more than previous one' by the rule 'two more than the previous one'. With it the counts '1, 2, 3, 4, 5, 6, —' shall be transforming as '2, 4, 6, 8, 10, 12, ——' .

Likewise instead of '1 as 2', there can be working with '2 as 1'. It shall be leading to the counts as ' ½ , , 1/6, 1/8, —'.

3. **Chase the text letter wise, in the sequence of their appearance in the text:**

UPSUTRA - 1

आनुरूप्ये

01	02	03	04	05	06	07	08	09	10
आ	न्	उ	र्	ऊ	प्	प्	य्	ए	अ

Total Letters	Vowels	Nasels	Consonents
10	5	-	5

Sub formulations of the text emerge as (1) आनु *anu*/to follow, (2) रूप्य, *Rupua*/form/frame and (3) येण, *aina/ anubanda* (bound)/limit.

In terms of these sub formulations the simple English rendering (resulting into working rule) comes to be : 'to follow the form/frame uptill the limit'.

4. **Working Rule:**

In terms of these sub formulations the simple English rendering (resulting into working rule) comes to be : 'to follow the form/frame uptill the limit',

The working rule of first Ganita Upsutra may be narrated as

(1) 'proportionately (as values in terms of artifices of numbers' and

(2) summitry (of forms along geometric formats within dimensional frames)

5. **Ganita Upsutra-1 presumes Ganita Sutra-1:**

'To follow the form – frame', presumes the existence of the form – frame, which is to be followed and this availability of the form – frame is there with the availability of Ganita Sutra-1.

6. **Letters ranges '16 and 10':**

The sixteen letters range of Ganita Sutra-1 focuses upon the values of '16' for the organization format of Ganita Sutra-1.

The artifice 16 admits re-organization as of 4 4 format.

This format is of spatial order; four steps long axes.

These four steps sequential chase is of four fold manifestation layers formats of four consecutive dimensional spaces/ four consecutive whole numbers.

The artifice '16', accordingly admits a split as a pair of four steps long chase formats, namely, (1) 0 + 1 + 2 + 3 and (2) (1 + 2 + 3 + 4).

The first chase range of steps 0 + 1 + 2 + 3 = 6 and second chase step range (1 + 2 + 3 + 4) = 10, with this split, would give us insight as to how, the second of these two chase steps ranges, lead to the organization format range of the ten letters text of Ganita Upsutra-1.

7. **Blissful exercise:**

It would be blissful exercise to permit the transcending mind to remain in deep sittings of trans and to be face to face with the two ranges split for artifice 16 as (*i*). (0,1, 2, 3) and (*ii*). (1, 2, 3, 4) and the second range (1, 2, 3, 4) being parallel to the range of Ganita Upsutra-1.

Further it would be blissful to have a pause and to be face to face with the above splits and the feature as that Ganita Sutra-1 takes to Ganita Upsutra-1.

Still further it also would be blissful to comprehend and to have an insight of the above features as that the sequential steps of each of two ranges, may it be

(0, 1, 2, 3) or (1, 2, 3, 4) are following the working rule of Ganita Sutra-1, that is, 'one more than before'.

Still further it would be very blissful to comprehend as that the transition and transformation from the first range (0, 1, 2, 3) to the second range (1, 2, 3, 4), there are transitions and transformations resulting into attainments for each of the four steps being as per the working of Ganita Sutra-1 itself as that the first step of the first range would transit into the first step of the second range by the working rule of Ganita Sutra-1 as that 0 + 1 = 1. And likewise would be the attainments for the second, third and fourth steps as well.

8. Two fold simultaneous sequential increase:

Let us have a fresh look at the two sequential ranges split for artifice 16 as under:

	First Step	Second step	Third step	Fourth step
First sequential range	0	1	2	3
Second sequential range	1	2	3	4

The transition from first step to second step of first sequential range is attainable in terms of working rule of Ganita Sutra-1 as that 0 + 1 = 1.

Likewise the transition from first step of first sequential range to the first step of second sequential range as well is attainable by the working rule of Ganita Sutra-1 as 0 + 1 = 1.

9. Two dimensional frame for sequential flow:

The above simultaneous pair of sequential flow steps, as such is of the format of two dimensional frame.

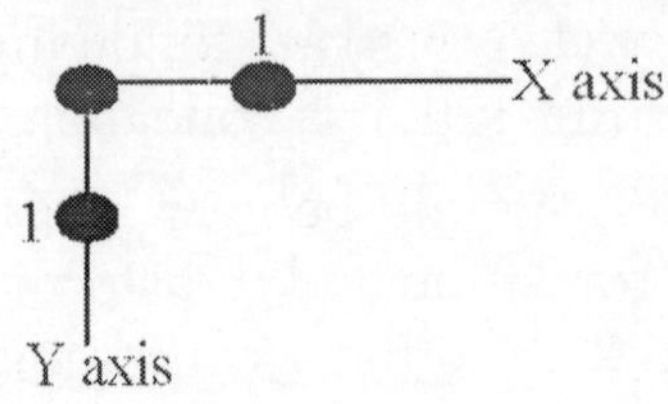

10. 4 4 matrix format:

0	1	2	3	= 6
1	2	3	4	= 10
2	3	4	5	= 14
3	4	5	6	= 18

Grand total of values= 48

Artifice 48 = 2 4 6

11. Manifestation layers organization along 4 4 format:

n–3	n–2	n–1	n	= 4n–6
n–2	n–1	n	n + 1	= 4n–2
n–1	n	n + 1	n + 2	= 4n + 2
n	n + 1	n + 2	n + 3	= 4n + 6
Grand total of values				= 16n

It would be blissful to note that for all values of n, the grand total of the values of all the sixteen placements of 4×4 format as above comes to be 16 n. It as such shall be bringing to focus the potentialities of the features of artifice '16' availed for the organization format of Ganita Sutra-1.

12. Sixth and first vowels:

It would further be very blissful to note that numerals 6 is at unit place and numeral 1 is at the ten value place for artifice '16'.

Parallel to it, it also would be very blissful to note that sixth vowel is 'ए' and the first vowel is 'अ'.

Further it would be very – very blissful to note that Lord Vishnu is the presiding deity of 6-space and as such its artifice value is '6'.

The ancient wisdom enlightenment, as preserved in Srimad Bhagwad Geeta, enlightens us as that Lord Krishna, the incarnation of Lord Vishnu, enlightens Arjuna as that 'amongst syllables, He 'Lord' is *Akara* (अ).

One may have a pause here and permit the transcending mind to remain in deep sittings of trans and to glimpse and to be face to face with the feature of organization formats of Ganita Sutras and Ganita Upsutras, as that sixth vowel is the first letter of Ganita Sutra-1 text while the first vowel (as syllable) is the first letter of Ganita Upsutra-1 text.

For satisfaction of the urge of the sadhkas as to the cyclic chase in reverse orientation as starting with the sixth vowel for the Sutras and starting with first vowel for the Upsutras, it is pointed with focus that the text of chapter-1 of Srimad Bhagwad Geeta as well as of the text of chapter-6 of Srimad Bhagwad Geeta is of same values (of 47 *shalokas* each).

This would help appreciate as to why the whole range of numbers are approached in terms of finite value place value system, like ten place value system in terms of only nine numerals and tenth place value and parallel to it, finite scales help chase the infinite ranges, and that is why 'INFINITE', is In Finite.

◆ ◆ ◆

66

GANITA SUTRA-2 AND GANITA UPSUTRA-2

1. Text of Ganita Sutra-2:

निखिलं नवतश्चरमं दशतः *Nikhilam Navatascramam Dasatah*	All from 9 and the last from ten

01	02	03	04	05	06	07	08	09	10
न्	इ	ख्	इ	ल्	अ	ं	न्	अ	व्
11	12	13	14	15	16	17	18	19	20
अ	त्	अ	श्	च्	अ	र्	अ	म्	अ
21	22	23	24	25	26	27	28		
-	द्	अ	श्	अ	त्	अ	:		

Total Letters	Vowels	Nasels	Consonents
28	12	3	13

2. Text of Ganita Upsutra-2:

शिष्यते शेषसंज्ञः *Sisyate Sesasamjnah*	That remains is remainder

01	02	03	04	05	06	07	08	09	10
श्	इ	प्	स्	अ	त्	ए	श्	ए	प्
11	**12**	**13**	**14**	**15**	**16**	**17**	**18**		
अं	स्	अं		ज्	ञ्	अ	ः		

Total Letters	Vowels	Nasels	Consonents
18	7	2	9

3. Transitions from Ganita Sutra-1 and Ganita Upsutra-1:

(*i*) For insight of the organization features of Ganita Sutras (and Ganita Upsutras) text, Ganita Sutra-1 be taken as the source sutra. From the organization format of Ganita Sutra-1, two fold transition be chased, firstly from Ganita Sutra-1 to Ganita Sutra-2

(*ii*) And secondly from Ganita Sutra-1 to Ganita Upsutra-1.

(*iii*) One way to reach at the transitions is to be in terms of the artifices values formats availed for Ganita Sutra-1 and Ganita Sutra-2, as well as for Ganita Sutra-1 and Ganita Upsutra-1.

(*iv*) It may be recapitulated as that sixteen letters range of the text of Ganita Sutra-1 avails organization format of artifice 16 as a pair of folds (four folds manifestation layers), respectively being (0, 1, 2, 3) and (1, 2, 3, 4).

(*v*) The two fold organization of organization format of artifice 16 as 0 + 1 + 2 + 3 = 6 and 1 + 2 + 3 + 4 = 10, as such, for its first fold of summation value 6 it shall be leading to the organization format artifice 28 of Ganita Sutra-2 and for its second fold of summation value 10 it shall be leading to the organization format artifice 10 of Ganita Upsutra-1.

(*vi*) Here, It would be relevant to note that artifice 6 is the artifice of first perfect number (6) and the artifice 28 is the artifice of the second perfect number (28).

(*vii*) As such the transition from Ganita Sutra-1 to Ganita Sutra-2 at its organization format values is of the features of transitions from first perfect number (6) to the second perfect number (28).

(*viii*) Further It would be relevant to note that artifice 6 = 1 + 2 + 3, as sum of its three proper divisors and artifice 28 = 1 + 2 + 4 + 7 + 14, as sum of its five proper divisors amounts to their inter linking of artifices 3 and 5 as of values of solid dimension (3-space as dimension) of transcendental domain (5-space as domain).

(*ix*) In the context, it also would be relevant to note that, the text of Ganita Sutra-1 is a composition of two formulations (एकाधिकेन पूर्वेण), of which the first formula-tion (एकाधिकेन) splits into a pair of sub formulations (एकः + अधिकेन), and thereby it being a composition of three formulations namely (एकः + अधिकेन + पूर्वेण).

(*x*) On the other hand the text of Ganita Sutra-2 is a composition of three formulations (निखिलं नवतश्चरमं दशतः), of which the second formulation (नवतश्चरमं) splits into three sub formulations (नवत + च + चरमं) and thereby it being the composition of five formulations (निखिलं नवत + च + चरमं दशतः).

(*xi*) The other way to approach this transition from Ganita Sutra-1 to Ganita Sutra-2 would be in terms of the compositions of the texts of these sutras. The text of Ganita Sutra-1 is a composition of two formulations, while the text of Ganita Sutra-1 is a composition of three formulations.

(*xii*) Still the other way to approach this transition would be in terms of the working rule of Ganita Sutra-1 itself.

(*xiii*) As far as the transition from Ganita Sutra-1 to Ganita Sutra-2 is concerned, the same as well, likewise can be of three fold approaches in terms of the artifices organization format, text composition and the working rule of Ganita Sutra-1 and of Ganita Upsutra-1.

4. Transition from Ganita Upsutra-1 to Ganita Upsutra-2:

(*i*) It would be blissful exercise to chase transition from Ganita Upsutra-1 to Ganita Upsutra-2.

(*ii*) This transition as well may be approached in multiple ways, of which the first may be in terms of artifices organization format values, namely '10' for Ganita Upsutra-1 and '16' for Ganita Upsutra-2.

(*iii*) One may have a pause here and have a fresh look at the inter linking of organization format values of Ganita Sutra-1 and of Ganita Upsutra-1 on the one hand, and of Ganita Sutra-1 and of Ganita Upsutra-1 on the other hand.

(*iv*) The inter linking of Ganita Sutra-1 and Ganita Upsutra-1 is of features of pairing of (16, 10) and the interlinking of Ganita Upsutra-1 with Ganita Upsutra-2 is of the features of pairing of (10, 16).

(*v*) It would be relevant to note that as there is two fold interlinking for Ganita Sutra-1 as being with Ganita Sutra-2 and with Ganita Upsutra-1; likewise there was two fold interlinking for Ganita Upsutra-1 with Ganita Upsutra-2 and with Ganita Sutra-1.

(*vi*) The other approach for transition from Ganita

Upsutra-1 to Ganita Sutra-1 and to Ganita Upsutra-2 shall be in terms of the text formulations of these Upsutras 1 and 2 and of Sutra-1.

(*vii*) It would be relevant to note that the text of Ganita Upsutra-1 is a composition of single formulation (आनुरूप्येण), while the text of Ganita Sutra-1 is a composition of pair of formulations (एकाधिकेन पूर्वेण). And the text of Ganita Upsutra-2 as well is a composition of pair of formulations (शिष्यते शेषसंज्ञ).

(*viii*) Still another approach for attaining transition from Ganita Upsutra-1 to Ganita Sutra-1 and Ganita Upsutra-2 would be in terms of the working rules of these Upsutras 1 and 2 and of Sutra-1.

(*ix*) It would be blissful exercise to approach from the symmetry rule (of proportionality) of Ganita Upsutra-1 to the 'one more than the previous one' working rule of Ganita Sutra-1. Here 'one' in particular, shall be unfolding its varied features, and the colourful spectrum of number systems unfolding from within 'one' itself under the symmetry frame for the working rule of one more than the previous one would follow.

(*x*) It would further be very blissful to approach from the symmetry rule of proportionality of Ganita Sutra-1 to asymmetry rule at the base of the working rule of Ganita Upsutra-2.

(*xi*) The working rule of Ganita Sutra-2 is of the features of enveloping asymmetry factor from the symmetry zone as by repeated subtractions of divisor from the dividend, the remainder is reached at, which when zero works out the dividend as fully divisible and hence dividend being the values of fully symmetrical zone, while the remainder is not zero, becomes of value lesser than that of divisor, and as such represents the asymmetrical value within the symmetrical zone being of multiples of the divisors.

(*xii*) The literal translations of the text of Ganita Upsutra-2 (शिष्यते शष्संज्ञ), as simple english rendering would emerge of the expression as that (शिष्यते) being 'that remains', and for the formulation 'शेषसंज्ञ' being 'that balance manifests (with a name as Noun)'. So the simple English rendering for 'शिष्यते शेषसंज्ञ' comes to be as 'that remains balance manifests as remainder'. It in more simple terms to mean 'that remains is remainder'

(*xiii*) One may have a pause here and may permit the transcending mind to remain in deep sittings of trans and to glimpse the way transition from the rule of 'remainders' of Ganita Upsutra-2, one can reach at the 'subtraction rule', at the base of the working methodology of Ganita Sutra-2 as 'All from nine and last from ten', which illustratively would be that to subtract 789 from 1000 would mean that 1000 be taken as 999 + 1 or as 99 (9 + 1). So to subtract 789 from 1000 would mean to subtract 7 from 9 and get 2, further to subtract 8 from 9 and to get 1 and as a last step to subtract 9 from 10 and get 1. The result there of all these steps would lead to 211 as result of subtraction of 789 from 1000.

5. Text as formulation of Sub formulations:

(*i*) Text of Ganita Sutra/Sub Sutra is a formulation.

(*ii*) Formulation permits splits into sub formulations.

(*iii*) Infact each letter of Vedic alphabet in itself is a formulation but for the purposes of compositions in terms of alphabet letters, these are accepted as such as of fixed unified frequencies states.

(*iv*) With it the knowledge systems chase of the texts

restrict itself up till the compositions reached at in terms of letters being the constituents which do not split further.

(*v*) Illustratively the text of Ganita Sutra-1 is a formulation 'एकाधिकेन पूर्वेण', which is a composition of two sub formulations 'एकाधिकेन' and 'पूर्वेण'.

The sub formulation 'एकाधिकेन' would further permit split as sub formulations 'एकः + अधिकेन'.

The sub formulation 'अधिकेन' would further permit split into sub formulations 'अधि and केन'.

(*vi*) The first range of formulations/sub formulations are the 'syllables'. Split of syllables, as such is not pursued as it would lead to letters.

(*vii*) Taking letters as these are, firstly syllables are composed, and then syllables need to further bigger formulations.

(*viii*) Taking that the split of text formulations as are the sub formulations availed by the text itself, as in case of Ganita Sutra-1, it avails a pair of sub formulations, namely 'एकाधिकेन' and 'पूर्वेण' and remaining restricted up till it, the artifices organization format beneath the text of Sutra shall be approached as it stands split by the artifices ranges formats of the formulations as in case of artifice sixteen is split as artifices nine and seven parallel to the letters availed respectively by the complete formulation text of Ganita Sutra-1 and letters availed individually by the pair of sub formulations of the text of Ganita Sutra-1.

(*ix*) In case of Ganita Sutra-1, the split/re-organization of sixteen as (9, 7) shall be providing us another way to approach the values and virtues of Ganita Sutra-1.

(*x*) It would be relevant to note that Brahman range (9-space/artifice 9) accepts unity state (7-space/

artifice 7) as its dimensional order. As such the coordination of (9, 7). One may have a pause here and permit the transcending mind to remain in deep sittings of trans and glimpse the values and virtues of the dimensional domains and to be face to face with artifice 9/9-space as the domain obeying the organization rule (one more than before) and the artifice 7/7-space as dimensional order of 7-space being the previous state of 9-space.

(*xi*) Likewise every Sutra text as a formulation shall be yielding approach for its features in terms of the formulation split to the level of the syllables.

(*xii*) As such, the conclusions comes to be that to have proper insight of the values and virtues range of the potentialities of Ganita Sutra, need would be to reach up till every syllable of the text.

(*xiii*) As the text of Ganita Sutras (including Upsutras), as well as the start with source reservoir syllable Om (ॐ) and the end fruit formulation Parnava being of 520 steps of letters availed, as such the syllables thereof shall be less than half of this range, as normally syllable has a consonant and a vowel, though in exceptional situation, vowels individually as well in terms of their inner folds express as syllables in themselves.

6. Initiation steps of learning:

Accordingly the initiation steps of learning in the above background come to be:

(*i*) Read the text.

(*ii*) Pronounce the each letter distinctively.

(*iii*) Chase every syllable with a pause.

(*iv*) Identify each sub formulation.

(*v*) Remain conscious of Sub formulations of every formulation (*vi*) Reach at the working rule of the formulation by natural chase of the meanings of sub formulations in the light of the syllables availed by the sub formulation.

(*vii*) Translate the working rule of the Sutra/Upsutra, under Sankhiya Nishtha/Systems.

(*viii*) Further translate the working rule of the Sutra/Upsutra under Yoga nishtha/systems.

(*ix*) Have insight about the transition bridges in the sequential order of the Sutras and Upsutras.

(*x*) Glimpse the cyclic systems of the system of working rules of Ganita Sutras.

One shall approach every aspect of Ganita Sutras by permitting the transcending mind to dive deep during prolonged sittings of trans to glimpse the processing processes of the systems of Ganita Sutras and to be face to face with it to imbibe their values fully by complete comprehension of them.

◆ ◆ ◆

67

GANITA SUTRA-3 AND GANITA SUTRA-4

1. Text of Ganita Sutra-3:

ऊर्ध्वतिर्यग्भ्याम् Urdhvatiryagbhyam
Vertically and crosswise

2. Sequential tabulation of the text:

01	02	03	04	05	06	07	08	09	10
ऊ	ध्	र्	व्	अ	त्	इ	र्	य्	अ
11	12	13	14	15					
ग्	भ्	य्	आ	म्					

Total Letters	Vowels	Nasels	Consonents
15	5	-	10

3. Sub formulation:

Text formulation 'ऊर्ध्वतिर्यग्भ्याम्' splits into a pair of sub formulations (1) 'ऊर्ध्व' and (2) 'तिर्यक' simple english rendering for these formulations amounts to : 'ऊर्ध्व' means upward (vertically upward) and 'तिर्यक्' means 'not straight'/of in between path/crosswise. As such the working rule becomes a composite rule of two steps namely ((1) processing vertically upward and (2) processing diagonal wise/crosswise).

4. Sankhiya format for the working rule:

This may be demonstrated with the help of two pairs of numbers, say (A, B) and (C, D) as under:

Step 1: To have placement for the first pair of numbers as

A B

Step 2: To have placement for the second pair of numbers, together with the first pair of numbers as follows

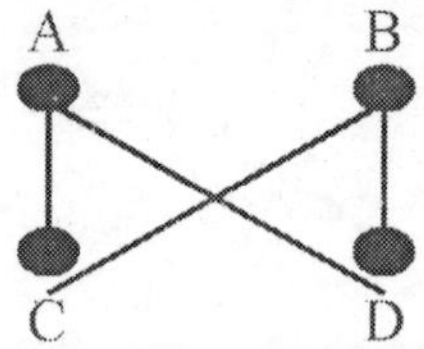

Step 3: A and C as well as D and B are vertically connected.

Step 4: C and B, as well D and A are diagonally/Crosswise connected.

Step 5: This format would be of the following features

Step 6: The internal depiction of cross stands accepted as symbol of multiplication operation ( ). Step 7 illustrative case of multiplication of numbers of pair of digits, say 12 and 13 amount to performance of the multiplication operation availing above format would be of following chase steps, as to be of three stages as follows:

First Stage: To multiply first digit of 12, i.e. 2 with the first digit of 13 i.e. 3 and to obtain the value 2 3 = 6

Second Stage: To multiply the second digit of 12 i.e 1 with the first digit of 13 i.e. 3 and to get the value 1 3 = 3 and to add the multiplication value of the first digit of 12 i.e. 2 with the second digit of 13 i.e. 1 which shall be yielding the value $2 \times 1 = 2$ with the end result being $3 + 2 = 5$, the middle place value for the multiplication of 12×13.

Third stage: To multiply second digit of 12 i.e. 1 with the second digit of 13 i.e. 1 and to obtain the third i.e. end place value being $1 \times 1 = 1$.

As such, the first digit of 12×13 comes to be 6. The second digit/middle digit of 12×13 comes to be 5 and the third i.e. last digit of 12 13 comes to be 1.

Conclusion $12 \times 13 = 156$.

This is straight application of the working rule of Ganita Sutra-3.

5. Recapitulation:

Predominantly along Sankhiya format, the working rule of Ganita Sutra-1 works out addition operation. The Ganita Sutra-2 works out the ten place value system. And the Ganita Sutra-3 works out the multiplication format.

6. Further recapitulation:

Ganita Sutra-1 avails line format for sequential increase of counting numbers/single digit numbers/numerals. Ganita Sutra-2 avails plane format for organization of numbers as double digit numbers as of placement value features. Ganita Sutra-3 attain transition from the pair of axes format to three axes format by attaining shift for the horizontal plane into a vertical plane.

7. To chase solids (space) in terms of pair of planes:

One may have a pause here and to have a fresh look at the working rule of Ganita Sutra-3 with a focus upon its pair of sub formulations processing steps to be vertical and to be crosswise.

With horizontal plane as a base, the vertical axes would

be the third axes of 3-space. The diagonal or parabolic curve path along vertical plane, as such shall be taking care of the processing rule of the second sub formulation of Ganita Sutra-3.

8. Text of Ganita Sutra-4:

परावर्त्य योजयेत् Paravartya Yojayet
Transpose and unite

9. Sequential tabulation of the text:

01	02	03	04	05	06	07	08	09	10
प्	अ	र्	आ	व्	अ	र्	त्	य्	अ
11	12	13	14	15	16	17			
य्	ओ	ज्	अ	य्	ए	त्			

Total Letters	Vowels	Nasels	Consonents
17	7	-	10

10. Transition from Ganita Sutra-3 to Ganita Sutra-4:

One may have a pause here and permit the transcending mind and to have a fresh look at the working rule of Ganita Sutra-3, particularly in respect of the pair of processing steps in terms of its pair of sub formulations.

Further one shall permit the transcending mind to glimpse the working rule of Ganita Sutra-4 being of two steps firstly 'to transpose' and secondly 'to unite' (with the set-up as it was prior to transposition).

It would be like uniting of object range with the image range as a reflection through a mirror.

11. 4-space format:

Sequential Ganita Sutra-4 as of fourth sequential step of the organization format of Ganita Sutras, in that sequence and order shall be making this fourth steps and hence format of Ganita Sutra -4 being of 4-space set-up features.

One may have a pause here and may recapitulate with focused attention as that 4-space is of a spatial dimensional order, that is here dimensional frame is going to be of four spatial dimensions.

12. Spatial dimension features:

Spatial dimension means 2-space/surface/plane playing the role of dimension. It as comparison to linear dimension/1-space in the role of dimension as of 3-space/line as axis, shall be making a shift from features of line format to that of features of surface format.

The surfaces admit pair of linear dimensions, which frame the surfaces/area as grid zones. The grid zones are of area measures while grid frame is of length measure. It is this basic distinction of the measures which make the basic difference of linear formats and of spatial formats.

For spatial dimensions/grid zones, there being a dimensional frame of pair of dimensions, as such transition from linear order 3-space to spatial order 4-space would mean that here to be at spatial set-up, need would be to be through grid zone itself to reach from first dimension to the second dimension of the spatial order itself.

One may have a pause here and to have a fresh look at the working rules of Ganita Sutra-3 and Ganita Sutra-4 and to be face to face with the features of the working rule of Ganita Sutra-4 in particular, whose focus is not only upon transposi-tion (reflection) to be at other half) but also upon the process-ing feature of uniting with that of the position of the set-up as it was prior to transposition. This, as such would mean that the first half as it was the position of the set-up prior to the transposition would get united with the second half emerging on transposition on and for the first half, and thereby the pair of halves unifying as a single integrated whole.

One may again have a pause here and to permit the transcending mind to remain in deep trans to glimpse this unification process under working rule of Ganita Sutra-4 and to imbibe these unification values to comprehend completely the transition process for transition from linear order set-up of single dimension to spatial dimensional order of a pair of dimensions.

13. **Further recapitulation:**

One shall again have a pause here and to further recapitulate the sequential steps of organization format of Ganita Sutras beginning from Ganita Sutra-1 and reaching up till Ganita Sutra-4.

As Artifice-4 admits re-organization as $2 + 2 = 2 \times 2 = -2 \times -2$, as such parallel features of 4-space shall be permitting absorption of the diversion features of reversal of orientations because of spatial dimensional order. It is in terms of it that the organization format of the Ganita Sutras as such shall be inherently in acquiring the cyclic features.

Because of the cyclic nature of the organization format of Ganita Sutra, a reverse chase from a format of Ganita Sutra-4 to the format of Ganita Sutra-1, as such shall be of the features of sequential chase as being of the values of four spatial dimensions, three spatial dimensions, two spatial and single spatial dimensions, which in the process shall be leading to Ganita Sutra-1 as being approachable in terms of single spatial dimension, because of which the sixteen letters organization range permitting split as a pair of manifestation layers namely (0, 1, 2, 3) and (1, 2, 3, 4) as formats of its pair of axes.

14. Multiple applied values of every Sutra:

It is because of the multiple features of the organization formats of Ganita Sutras, collectively as well as individually that every Sutra has multiple applied values and it is because of it that the single Sutra working rule makes it possible to cover different arithmetic and other mathematical operations.

◆ ◆ ◆

68

GANITA SUTRA-5 AND GANITA SUTRA-6

1. Text of Ganita Sutra-5:

शून्यं साम्समुच्चये *Sunyam Samyasamuccaye*
If the Samuccya is the same it is zero

2. Sequential tabulation of the Ganita Sutra-5 Text:

01	02	03	04	05	06	07	08	09	10
श्	ऊ	न्	य्	अ	ं	स्	आ	म्	य्
11	12	13	14	15	16	17	18	19	20
अ	स्	अ	म्	उ	च्	च्	अ	य्	ए

Total Letters	Vowels	Nasels	Consonents
20	8	1	11

3. Text of Ganita Sutra-6:

(आनुरूपये) शून्मन्यत् (Anurupye) Sunyamanyat
If one is in Ratio the others is Zero

4. Sequential tabulation of the Ganita Sutra-5 Text:

01	02	03	04	05	06	07	08	09	10
आ	न्	उ	र्	ऊ	प्	य्	ए	श्	ऊ
11	12	13	14	15	16	17	18	19	
न्	य्	अ	म्	अ	न्	य्	अ	त्	

Total Letters	Vowels	Nasels	Consonents
19	8	-	11

5. Approaching pair of Sutras at a time:

(*i*) One way to approach Ganita Sutras is Sutras wise in the sequence and order of the Ganita Sutras.

(*ii*) This approach shall be focusing also upon the features of the sequential order of the Sutras because of which Ganita Sutra-1 shall be acquiring features of artifice-1 and parallel to it of 1-space, as well as of its representative regular body, namely 'interval'. Likewise Ganita Sutra-2 shall also be acquiring features of artifice 2 and of 2-space, as well as of square/circle as its representative regular bodies. And so on would be the acquisition of the Ganita Sutras-3, 4 and ahead.

(*iii*) The other way to approach Ganita Sutras is as a pair of consecutive Ganita Sutras. In terms of it Ganita Sutras-1 and 2 would permit simultaneous approach. Likewise Ganita Sutras-2 and 3, and ahead Ganita Sutras-3 and 4, Ganita Sutras-4 and 5, and Ganita Sutras 5 and 6 shall permit simultaneous approach.

(*iv*) This as such shall amount to 1-space and 2-space, 2-space and 3-space, and likewise would be the formats for simultaneous approach for the pair of Sutras being approached simultaneously.

(*v*) Here it would be relevant to note that N-space and

(N + 1) space are interconnected as boundary and domain of hyper cube (N + 1).

(*vi*) With it, square as hyper cube 2 as 2-space/surface area enveloped within linear boundary shall be providing a format for simultaneous approach to Ganita Sutra-1 and Ganita Sutra-2.

(*vii*) A step ahead cube as hyper cube 3, as solid enveloped within surface, shall be providing a format for simultaneous approach for Ganita Sutra-2 and Ganita Sutra-3.

(*viii*) Likewise, Ganita Sutras-5 and 6 can be simultaneously approach by availing the format of hyper cube 6 (6-space enveloped within 5-space).

6. ***Vyakata, Avakata, Avakato-avakatat and Pursha:***

Ancient wisdom enlightens as that Vyakata, Avakata, Avakato-avakatat and Pursha are in sequential order.

Vyakata is expressed/macro state (3-space), *Avakata* is micro state (4-space), *Avakato avakatat* is casual state (5-space) and a step ahead is *Pursha* format (6-space).

7. **Casual state (5-space) and Step ahead (6-space):**

Casual state is a zero value state in reference to the Macro state. And pursha State, which is even step ahead, as well would be of zero value state in reference to micro state.

One may have a pause here and have a fresh look at the text and working rules of Ganita Sutras-5 and 6 and to comprehend and imbibe the pair of states of zero value.

8. **Ganita Sutras-5 and 6:**

Ganita Sutra-5 working rule 'If the Samuccya is the same it is zero' and Ganita Sutra-6 working rule 'If one is in Ratio the others is Zero' account for and avail the pair of features of zero value state.

To appreciate working rule of Ganita Sutra-5, one can pick up equal weight pebbles in hands and through them up. If, both stones reach up till same heights then it can be said that both stones have been thrown up with equal force.

To appreciate the working rule of Ganita Sutra-6 'If one is in Ratio the others is zero', one shall Comprehend as that AB = 0 would imply that for non zero A, B would be Zero.

9. Recapitulation:

Ganita Sutra-1 chases counting infinity step by step, in as large number of steps, as need be. Ganita Sutra-2 transits from single digit counts to double digit counts by transiting to place value system for organization of counts (number).

Ganita Sutra-3 avails the one half of the vertical plane while Ganita Sutra-4 focuses upon the second half of the vertical plane by having it united with the first half of the vertical plane. Now, Ganita Sutra-5 avails equality format for reaching at zero value. It is, in a way the subtraction of an equal and reaching at zero value. It is expressible as A – B = 0 also as that A = B further as that XA = XB. On the other hand Ganita Sutra-6 avails the division feature. If AB = 0 would imply that for non zero A, B would be Zero.

10. Blissful Exercise:

It would be blissful to permit the transcending mind to remain in prolonged sittings of deep trans to glimpse the features of 'Zero', as absolute value, being of affine state and different then that of the features of 'cipher'/small.

One may blissfully chase '0 + 0 = 0', '0 – 0 = 0', '0 0 = 0', 0-space, 0-space content, hyper cube-0,

0-space playing the role of dimension of 2-space, 0-space plays the role of dimension of dimension of 6-space, 0-space plays the role of origin of (–1-space), 0-space plays the role of boundary of 1-space and further as that Different roles of Zero along 4 4 format of values as follows:

–3	–2	–1	0	Value = – 6
–2	–1	0	1	Value = – 2
–1	0	1	2	Value = + 2
0	1	2	3	Value = + 6
	Grand value			'0'

Still further it would be blissful to comprehend as that 0/ 0 is not definable. Still further as that + 0 is not different then – 0.

11. Sustenance of *Pursha* format:

It would be very blissful to permit the transcending mind to dive deep to the level of dimension of dimension of the creator space (4-space) to face to face with the sustenance level of our existence phenomenon/along the Pursha format.

5-space is at the base of the origin fold of manifested creations. As such transcendence at the origin fold is a pilgrimage in casual state, a step ahead of the micro state existence phenomenon. Still a step ahead to the self-referral 6-space is the Pursha format. As such, the zero value state/0-space would be in the role of dimension of dimension of dimension of 6-space/ Pursha format and hence the regulation of human existence phenomenon of Pursha format goes for its sustenance at the level of dimension of dimension of the creator space (4-space).

◆ ◆ ◆

69

GANITA SUTRA-7

1. Text of Ganita Sutra-7:

संकलव्यवकलनाभ्याम्
Sankalana-vyavakalanbhyam
By addition and by subtraction

2. Sequential organization of the text of Ganita Sutra-7:

01	02	03	04	05	06	07	08	09	10
स्	अ		क्	अ	ल्	अ	न्	आ	व्
11	12	13	14	15	16	17	18	19	20
य्	अ	व्	अ	क्	अ	ल्	अ	ल्	आ
21	22	23	24						
भ्	य्	आ	म्						

Total Letters	Vowels	Nasels	Consonents
24	10	1	13

3. '1 as 2':

'1 as 2', essentially means '1' as a pair of halves (½, ½).

This feature as such helps approach '1', in terms of '½'.

4. **'2 as 1':**

'2 as 1' essentially means to club 'two entities' as single 'paired' entity. One of the features of it also would be to approach '2' as '1, 1', which for sum of its aspects shall be of the format of '1 as 2'.

5. **'1 as 2' and '2 as 1':**

The simultaneous approach of '1 as 2' and '2 as 1', shall be of distinct features than that of '1 as 2', as well as of '2 as 1'. As 0 + 1 =1, as such along the format of a line there may simultaneously exists points as well as intervals/segments of line.

Likewise as '$2 = 0 + 1 + 1 \times 1$', as such within the surface, simultaneous would exists points, lines and surfaces.

This, in a way shall be amounting to approaching '1 as sequential range (0 + 1), and ahead '2' as the sequential range (0 + 1 + 2) and so on.

The split of surface (area) unit 1^2 as 1^1 1^1 as approach to area in terms of chase along a pair of axes; and further approach to surface/area unit 1^2 as ($1^2/2$, $1^2/2$) shall be of the format $(1\times 1)/(1 + 1)$, shall be bringing to focus different features of area unit from the windows of 'length units'.

However, line, itself of pair of orientations. Along these pair of orientations, a pair of length units would be available. It is because of reflection pairing of the pair of orientations, the corresponding length units as well would get paired as (+1, -1).

Therefore this simultaneous chase of '1 as 2' and '2 as 1', shall be also bringing into operation the feature of reversal of orientations as well. It is this reversal of orientation features, would add absolutely new features and would provide a qualitative transition and transformation over the features of '1 as 2' and '2 as 1' being operating individually even though both of these operations come into play for any given set-up.

6. **Transition from 1^3 to 2^3:**

To comprehend the transition and transformation from the pair of operations '1 as 2' and '2 as 1' coming into play individually, even one after the other over the same set-up, 'both of these operation' '1 as 2 and 2 as 1' coming into play simultaneously, as an integrated single operation, may be appreciated by having a fresh look at '1^{3}'/ cube and '2^{3}'/cube permitting split into synthesis of eight cubes of the values '1^{3}'/cube. As a concrete exercise, one can try to put eight cubes together to have another cube. And, simultaneously, one shall cut a given cube into eight cubes. In both processes one shall be coming face to face with the situation that while in the first instance of getting eight cubes together as a single cube, as that it shall be having eight corner points, twelve edges and six surfaces for each cube and this set of 64 corner points, 96 edges and 48 surfaces are to result into one cube of eight corner points, twelve edges and six surfaces for eight units of volummes. Here it would be relevant to note that though the volumme of eight cubes and of the single cube being of eight units remains the same but for all other components. There is going to be a big variance.

Now approaching otherwise and starting with a single cube of eight corner points, twelve edges and six surfaces enveloping unit volumme, into eight sub cubes, it would be coming to focus as that the total corner points, edges and surfaces for these eight sub cubes would be as follows:

First sub cube shall be a complete cube of 1/8 volumme but it shall be having only one corner points and three half edges and three quarter unit surfaces. Likewise would be the structural set-up for all other seven sub cubes as well. These set-ups together shall be providing a split for 1^3 as 8 $1^3/8$, however none of these eight sub cubes shall be a complete cube

7. **Three dimensional frame of half dimensions:**
One shall have a pause here and to have a fresh look at the structural set-ups of eight sub cubes of a cube it would bring to focus that the available corner points is the origin point of a three dimensional frame of half dimensions. As such the glaring feature coming to focus for the split of cube into eight sub cubes is that the centre of the cube/origin of 3-space is a seat of hyper cube 4 of solid boundary of eight components and as such the split of cube as eight sub cubes, as above is there because of eight solid components set-up of the boundary of hyper cube 4, the representative regular body of 4-space.

8. **Transition from linear dimensional order to spatial dimensional order:**
One may again have a pause here and have a fresh look at the set-ups of eight sub cubes of cube and of solid boundary of eight components of hyper cube 4. it would be coming to focus as that it is because of the spatial dimensional order of 4-space that the linear order of 3-space permits split of a three dimensional frame into three dimensional frames of half dimensions. It is in the process that cube splits into sub cubes within three dimensional frames of half dimensions. The structural set-up of these eight sub cubes being of one corner point, three half edges and three half surfaces for $1/8^{th}$ volume unit and that the corner point and three half edges together synthesizing as a functional model of a three dimensional frame of half dimensions for each of these eight sub cubes further brings to focus as that these eight three dimensional frames of half dimensions are coordinated as four sets of two three dimensional frames of half dimensions with dimensions being in opposite orientations.

9. Four sets of paired three dimensional frames of half dimensions:

The availability of four sets of paired three dimensional frames of half dimensions and as that the axis of such paired dimensional frames being of opposite orientations and further as that the six surfaces of paired dimensional frames together being of the surface area equal to half of the surface area of the unit cube providing split for such dimensional frames, would bring to focus the features of four dimensional frame of spatial dimensions.

One may further have a pause here and permit the transcending mind to remain in deep sittings of trans and to glimpse this phenomenon of four dimensional frame of four spatial dimensions coming into play and coordinating four sets of paired three dimensional frames of half dimensions.

Transition from artifice 7 to artifice 8. The insight about the above split of cube into eight sub cubes and imbibing of its value shall be helping to chase transition from artifice 7 to artifice 8.

The artifice 7/number 7/numeral 7 is the biggest prime numeral of the ten place value system. The artifice 8 permits re-organization as 2^3 .

As such the transition from artifice 7 to artifice 8 shall be inheriting the transition features as of transition from the set-up of 1^3 to 2^3 of it the central feature would be of split for the dimension as a pair of half dimensions, and also coordination for the half dimensions of opposite orientations.

It is this coordination for the dimension format as of a pair of half dimensions but of opposite orientations deserves to be imbibed fully. It would be relevant to note that when we draw an axes XOY, with O as origin and OX and OY as pair of half axes, we infact are availing here the format of coordination of pair of half dimensions but of opposite orientation along the format of a full axes.

It would be like availing simultaneously a pair of opposite orientations like that of OX and OY. It is parallel to having positive length unit along one orientation and negative length unit along the other orientations.

10. **Working Rule of Ganita Sutra-7:**

11. **Simple English rendering for the working rule of Ganita Sutra-7** is by addition and subtraction. It is to avail the format of pair of orientations of a line. It is to go by addition and by subtraction individually as well as simultaneously.

◆ ◆ ◆

70

GANITA SUTRA-7 AND GANITA SUTRA-8

1. Text of Ganita Sutra-8:

पूरणापूरणाभ्याम् Puranapuranabhyan
By the completion or non-completion

2. Sequential organization of the text of Ganita Sutra-8:

01	02	03	04	05	06	07	08	09	10
प्	ऊ	र्	अ	ण्	आ	प्	ऊ	र्	अ
11	12	13	14	15	16				
ण्	आ	भ्	य्	आ	म्				

Total Letters	Vowels	Nasels	Consonents
16	7	-	9

3. Mid-way of Artifice 7 and artifice 8:

Mid-way of artifice 7 and artifice 8 as to be of the value '7 and ½ ' like middle point of linear unit/centre of units surface or of solid unit/origin of hyper cubes/ spaces, would be like middle points/centre/origin of dimensional axes. It shall be of the format of a coordination of half dimensions of opposite orientations.

The simultaneous play of '1 as 2' and '2 as 1' shall be making the seat of the middle point/centre/origin of the dimensional axes format as to be of transitional features for linear order transiting into spatial order and other way round as well permitting transition from spatial order into linear order.

It would be like that of a format where $+ 0 = - 0$; and further as that $0 + 0 = 0, 0 \quad 0 = 0$. It would be a blissful exercise to have comparative comprehension of these features of zero value with the features of artifice 4 as that $2 + 2 = 4, 2 \times 2 = 4$ and $(-2) \times (-2) = 4$.

It would be relevant to note that 0-space plays the role of dimension of 2-space and ahead 2-space plays the role of dimension of 4-space. Further it also would be relevant to note that the centre/origin point of cube even at zero value is along the 4-space seat for it.

One may have a pause here and permit the transcending mind to remain in deep sittings of trans and to glimpse this phenomenon of features of zero value as well as of artifice 2 value.

With prolonged sittings of trans one shall be acquiring proper insight of this phenomenon being responsible for the coordination of pair of half dimensions of opposite orientation at the middle/centre/origin.

Phenomenon of hyper circles being of increasing sequential values for hyper circle 1 to 7 and that the same starting transiting into decreasing sequential values from hyper circle 8 onwards. However, infact this transition takes place the mid-way of artifice 7 and artifice 8

4. Split for a domain:

Let us have pause here and to have a fresh look at the set-up of a closed interval (the interval whose both end points are intact. Its split at the middle or at any point in between the pair of end points, it shall be bringing to focus as that the said point, on split shall be part of either of the two portions of the domain (here, closed interval).

For pointed attention, taking the end points of the closed interval being A and B respectively, and the reference point, say 'O' at which split is going to be had for the domain of closed interval [AB], on its split shall be having its two parts, namely [AO], where the point 'O' is of this part, and the second part being

(OB], as that the point 'O' is not part of it. With it, the portion [AO] would be a closed interval while the portion (OB] shall be a half closed interval.

The other way round, if the reference point 'O' would be included in the second part, then the split of closed interval [AB] shall be a pair of parts, namely half closed interval [AO) and closed interval [OB].

It would be relevant to note that, in both cases, the split for closed interval is always into a pair of intervals, one of them being a closed interval and other being a half closed interval.

One may have a pause here and permit the transcending mind to comprehend the values of this split for the closed interval into a closed interval and half closed interval. The structure of the start with closed interval is carried forward by the closed interval part thereof while the remaining part remains only half closed interval.

It is because of this feature that, the closed interval is taken to be a complete in its set-up while the half closed interval remains in complete because of the want of one of its ends being missing.

5. **Working rule of Ganita Sutra-8:**

   In the light of the above background, it would be blissful to chase the working rule of Ganita Sutra-8 as being 'by completion and by non completion' working rule of Ganita Sutra-8 along the above split format for the close interval.

6. **Ganita Sutra-7 and Ganita Sutra-8:**

   One may have a pause here and to have fresh look at the formats of the working rules of Ganita Sutra-7 and Ganita Sutra-8. It would be blissful to note that in case of Ganita Sutra-7 as well as in case of Ganita Sutra-8, the features of '1 as 2' come into play, as in case of Ganita Sutra-7 pair of opposite orientations are being taken care of and in case of Ganita Sutra-8 closed interval and half closed interval as well are being simultaneously taken care of.

7. **Transition from Ganita Sutra-7 to Ganita Sutra-8:**

   The transition from Ganita Sutra-7 to Ganita Sutra-8 may be had in many ways by availing different features of the values of artifice 7 and artifice 8.

   One way to approach this transition would be as that 3-space is of seven geometries range parallel to which cube shall be having seven distinct versions in terms of the presence or absence of surfaces of the cube. However 3-space accepts split into eight octants and parallel to it cube admits splits into eight sub cubes. It would be parallel to the solid boundary of 4-space/ hyper cube 4 being of eight components.

   With it, the transition from the format of Ganita Sutra-7 to that of Ganita Sutra-8 would be there because of the spatial boundary of cube on the one hand and solid boundary of hyper cube 4 on the other hand.

8. **External and internal splits:**

   One way to approach the transition from artifice 7 to artifice 8 and parallel to it from Ganita Sutra-7 to Ganita Sutra-8 shall be in terms of the external/ boundary of cube and internal/hyper cube 4 seat for origin of cube.

9. **Transition from linear order to spatial order:**

   As the cube domain being of a linear order set-up and its origin being of a spatial order set-up, as such the transition from Ganita Sutra-7 to Ganita Sutra-8 would turn out to be a transition from linear dimensional order to spatial dimensional order.

10. **Complete and incomplete domains:**

    Perfect squares, perfect cubes and perfect higher powers lead to complete domains. on the other hand, imperfect squares, imperfect cube and imperfect higher powers shall be leading to imperfect domains.

    Accordingly there shall be a pair of formats for chasing the values as along complete domain as along incomplete domains.

11. **Complete boundary and incomplete boundary:**

    The completeness or otherwise of the boundary makes the domains being complete or incomplete. And interval with both end points intact shall be a complete domain as its boundary is complete with both ends points being intact.

    Likewise square with its all four boundary lines being intact shall be making a complete domain because of completeness of its boundary. Otherwise in case of any of its corner points or any of its boundary lines being missing, it shall be making it an incomplete square domain. Likewise would be the position in respect of cubes and hyper cubes.

12. Complete domains of a space:

Every closed interval is a complete domain of 1-space. Further as every closed interval being of infinite range of points, as such they carry identical structural features.

Likewise every complete domain/square carries same structural set-ups. So would be the position in case of complete cubes as well as the complete hyper cubes of a given space.

◆ ◆ ◆

71
GANITA SUTRA-9 AND GANITA SUTRA-10

1. Text of Ganita Sutra-9:

चलनकलनाभ्याम् *Calana-kalanabhyam*
Differentiation-integral Calculus

2. Sequential organization of the text of Ganita Sutra-9:

01	02	03	04	05	06	07	08	09	10
च्	अ	ल्	अ	न्	अ	क्	अ	ल्	अ
11	12	13	14	15	16				
न्	आ	भ्	य्	भ्	म्				

Total Letters	Vowels	Nasels	Consonents
16	7	-	9

3. Text of Ganita Sutra-10:

यावदूनम् *Yavadunam*
By deficiency (of double of it)

4. Sequential organization of the text of Ganita Sutra-10:

01	02	03	04	05	06	07	08	09
य्	आ	व्	अ	द्	ऊ	न्	अ	म्

Total Letters	Vowels	Nasels	Consonents
9	4	-	5

5. Recapitulation and beginning afresh:

Before proceeding one may have a pause to have recapitulation of the steps so far to have a fresh beginning in the light of the features of the organization format of Ganita Sutra-1 and Ganita Upsutra-1.

The organization range of the Ganita Sutra-1 is sixteen letters long while that of Ganita Sutra-1 is only ten letters long. The Ganita Sutra-1 is a composition of two formulations of nine letters and seven letters ranges respectively while Ganita Upsutra-1 is a composition of single formulation of ten letters. The split of artifice 16 as 6 + 10 together with the re-organizations for artifices 6 and 10 being (0, 1, 2, 3) and (1, 2, 3, 4) as of four fold manifestation formats, and these together as a split of pair of manifestation layers for the five fold transcendence range (0, 1, 2, 3, 4) shall be giving us insight as to how the sixteen Ganita Sutras range may deserve to be approached as Ganita Sutras-1 to 6 and as Ganita Sutras 7 to 16.

Let us have a fresh look at the features emerging so far and as that the re-organization of 16 as 6 + 10, and as that Ganita Sutra-1 avails artifice 10, would help us have an insight as that Ganita Upsutra-1 because of its range of artifice 10 becomes the Upsutra (Upsutra-1) of Sutra-1, being parallel to the second sub range of Ganita Sutra-1.

6. Sankhiya Nishtha and Yoga Nishtha:

The re-organization of artifice 16 as 6 + 10 and parallel to it the pair of manifestation layers (0, 1, 2, 3) and (1,

2, 3, 4) shall be a split of transcendence range (0, 1, 2, 3, 4) into a pair of manifestation layers (0, 1, 2, 3) and (1, 2, 3, 4). This feature would help have an insight as to the *Sankhiya Nishtha and Yoga Nishtha* running parallel to each other as complementary and supplementary phenomenon.

Here it also would be relevant to note that, it is because of this feature that the Ganita Sutras and Upsutras ranges of values sixteen and thirteen as such run parallel to 5 + 6 + 5 and 4 + 5 + 4 formats. Here one may have a pause and have a fresh look at the features of hyper cube 5 whose origin fold is 6-space. Likewise the origin fold of hyper cube 4 is 5-space. Further as that 4-space plays the role of boundary of hyper cube 5/5-space as domain.

It is this coordination of 4-space as boundary fold of 5-space domain, and the fact that 4-space is a spatial order set-up and 5-space is a solid order set-up and further as that the origin seat within domain fold is of dual features as of domain fold as well as of origin fold, and as such in the context of domain fold it is to be of higher value and as such the re-organization for the domain fold together with its origin, in respect of hyper cube 5 being 5 + 6 + 5 and in respect of 4-space being 4 + 5 + 4.

One may again have a pause here and have a fresh look at the set-up of hyper cube 5 and to comprehend as to how the boundary fold and domain fold roles being played by 4-space and 5-space respectively have simultaneous manifestation because of the 'पूरणापूरणाभ्याम्' /'Puranapuranabhyan' (By the completion or non-completion). This together with the rule of Ganita Sutra-7 'संकलनाव्यवकलनाभ्याम्' / 'Sankalana-vyavakalanbhyam' (By addition and by subtraction), chasing simultaneously addition and subtraction will provide an insight for the simultaneous manifestation of four consecutive dimensional contents as hyper cubes and as a reverse process as to how their individual folds as well deserves to be chased as independent manifestations of four consecutive dimensional spaces and thereby emerging 4 x 4 matrix format for their for the complete chase of the hyper cubes.

This, this way makes the artifice value 16 availed for the range of sixteen Ganita Sutras is to simultaneously work out the parallel systems of *Sankhiya Nishtha and Yoga Nishtha*.

7. **Transition from four folds set-up to five fold set-up:**

   To have an idea of transition from four folds set-ups to five folds set-ups, the first feature to be taken note of that five points are coordinated by four units. Secondly as that 5×5 matrix format shall be permitting super imposition there upon a 4×4 matrix format. With its corners getting super imposed upon the centres of $5 \times 5 = 25$ zones of 5   5 matrix grid. To transcend from super imposed 4×4 (16) grid zones to base 5×5 (25) grid zones is a process of *Yoga Nishtha*. Here It would be relevant to note that 4×4 matrix format shall be having super imposed 3×3 matrix format and these $3 \times 3 = 9$ flow streams shall be fulfilling $5 \times 5 = 25$ grid zones of the base grid like 3-space playing the role of dimensional order of 5-space within creator space (4-space).

8. **Transition from Ganita Sutra-8 to Ganita Sutra-9:**

   Ganita Sutra -8 'पूरणापूरणाभ्याम्' / '*Puranapuranabhyan*' (By the completion or non-completion) avails the format of split of domain folds as of complete domain part and of an incomplete domain part. The distinguishing feature of complete and incomplete domains (parts) shall be as to the difference of a single point (in case of 1-space domain), single line (because of 2-space domain), single surface (in case of 3-space domain) and solids/hyper solids (in case of hyper solids 4 onwards). These points/ lines/surfaces/solids/hyper solids themselves being manifestation layers/hyper solids, as such shall be of many features. Of these one feature is as that a point is a boundary point of a line, line is a boundary line of surface, surface is boundary surface of solid, solid is boundary solid of hyper solid and so on Hyper solid (n – 1) is boundary hyper solid of hyper solid (n).

Further as that line has two boundary points, surface has four boundary lines, solids have six boundary surfaces, hyper solid –4 has eight boundary solids and so on, in general hyper solid n has 2n hyper solid (n – 1) boundary components.

As such the complete and incomplete domain split takes place just in terms of half of the boundary (components/ hyper cubes at boundary). However the complete boundary, naturally would be of double number of boundary components (hyper solids).

Therefore the transition from Ganita Sutra-8 to Ganita Sutra-9 shall be to reach at the complete stripping off of the domain of its boundary and in the reverse process to have a full enveloping of the domain (in terms of its full boundary).

9. **Transition from Ganita Sutras 8 and 9 to Ganita Sutras-10:**

Ganita Sutra-8 focuses upon the completeness and incompleteness of the domains which ultimately may turn out to be a single point in case of line, and a single line in case of a square and so on. On the other hand, Ganita Sutra-9 provides a process for reaching from the difference gap of complete domain and incomplete domain as to be of the size of the range of full boundary, which in case of lines comes to be two points in case of line/four lines in case of square, six surfaces in case of cube, eight cubes in case of hyper cube 4 and so on (2n hyper cubes in case of hyper cube (n + 1)).

The minimum deficiency between complete and incomplete domains may be of a single point/half boundary, and the maximum difference between the complete and incomplete domain shall be of two points/ boundary. As in case of line as domain, the difference between complete and incomplete such domains is of the range of single point (as half boundary) and of pair of points (as full boundary), as such the focus in general is to be upon the half boundary deficiency and full boundary deficiency in terms of the working rule of Ganita Sutra-10.

10. First consonants ka 'क्':

Ancient dictionaries associate meanings with the first consonant 'क्' as Brahma/four head lord Creator the supreme and the Shiv, the five head transcendental Lord. The idols of Lord Brahma and Lord Shiv, respectively are of the formats of hyper cube 4 and hyper cube 5. The hyper cube 4 domain manifests four fold idol format while the hyper cube 5 domain provides five fold transcendence range.

The four fold manifestation range of creator space (4-space) is of values (1) रीयि्/physical matter/3-space content (2) माया/micro state content (4-space content (3) लीला/casual state content (5-space content) and (4) प्राण/life breadth/self-referral state content (6-space content). As such the four folds together are of expression format of letters र, म्, ल्, ह्,/3-space, 4-space, 5-space, 6-space. The 4×4 matrix format shall be of values

3	4	5	6
4	5	6	7
5	6	7	8
6	7	8	9

3-space	4-space	5-space	6-space
4-space	5-space	6-space	7-space
5-space	6-space	7-space	8-space
6-space	7-space	8-space	9-space

The 5×5 transcendental base for 4×4 manifestation available shall be of the values

3	4	5	6	7
4	5	6	7	8
5	6	7	8	9
6	7	8	9	10
3-space	4-space	5-space	6-space	7-space
4-space	5-space	6-space	7-space	8-space
5-space	6-space	7-space	8-space	9-space
6-space	7-space	8-space	9-space	10-space

11. Consonant 'ल्' and formulation 'कल':

The basic formulation 'कल' avails the letters 'क्' and 'ल्', the meanings of 'कल' are 'yesterday', 'future' and 'machine'. Machine also means 'mundane'/static. Against it is the formulation 'चल' which means 'motion'. With it the formulations 'सकलन, व्याकलन, कलन' And 'चलन' acquire additional features which shall be at the base of working rules of Ganita Sutras-7 and 9 in particular.

12. Present day Disciplines of Calculas:

The present day Disciplines of Calculus because of its focus upon half boundary is responsible for its paradox situation like 'every where continuous functions but no where permitting differentials'. Further as may be seen the very limited scope of such rule, the fresh view need be taken to enlarge the functional domain of such processes for sequential reach from domain to boundary of hyper cubes, the functional values of format of Ganita Sutra-9 in particular deserves to be availed.

13. Test of the set-up of cube:

The test of the success of the calculas approach would be the reach at the structural set-up of cube, particularly that of cube as being the starting point.

Let us have a pause here and permit the transcending mind to have a fresh look at the geometric envelope stitched for 3-space content/volumme of cube, representative regular body of 3-space. It has eight corner points, twelve edges and six surfaces for a unit volumme. These artifices together namely eight, twelve, six, one, sum up $8 + 6 + 12 + 1 = 27 = 3^3$. The focus here upon artifice 3. 3-space, third degree, three dimensions further as that $3 = 1 + 1 + 1$, it that way gives a focus upon artifice 1/linear dimension/1-space.

Here it would be relevant to note that $3 = 1 + 2$. Let us have a fresh look at interval/representative regular body of 1-space as a unit length and two end points. These components together are of artifice value $(1 + 2 = 3)$.

Further, let us have expression for $(A^1 + 2A^2)^3 = A^3 \ A^0 + 3A^2 (A^0)^2 + 3A^1 (A^0)^3 + (A^3)^0 \times (A^0)^3 = 1A^3 + 6A^2 + 12 A^1 + 8 A^0$. The sum of the components are $1 + 6 + 12 + 8 = 27$ parallel to the components of geometric envelope of cube stitched for unit cube value in terms of eight corner points, twelve edges and six surfaces.

Further here also would be relevant to note that $[A^1 + 2A^0]^1 = 1A^1 + 2A^0$ as of components $1 + 2$ parallel to the geometric envelope for unit length stitched in terms of two points. And ahead $[A^1 + 2A^0]^2 = 1A^2 + 4A^1 + 4A^0$ as of components $1 + 4 + 4$ parallel to geometric envelope for unit area stitched with four corner points and four boundary lines.

Now let us have a fresh look at hyper cube 3. it shall be a 3-space content lumb/volumme bagged in 2-space content boundary of 6 surfaces/components to reach from it to the boundary of six surfaces as of 24 lines, as a second step being boundary of boundary of 3-space, it shall be bringing into play a dimensional frame for the second time and as such when it is accounted for to make it within a single frame, it shall lead to 24/2 = 12 lines remaining available. And at the next step, that is at third step, the repetition of the process shall be as that the reach for it is from double frame to third frame and as such 12×2 = 24 points would get reduced to 24/3 = 8 points.

15. **Recapitulation and Conclusion:**

16. **The Recapitulation and Conclusion** from above is that the calculas with its repeated application is to appropriately take care of the sequential steps of dimensional frames being availed. Likewise whole range of hyper cubes can be stripped off from their geometric envelopes/boundary as well as that every dimensional content can be put into properly stitched geometric bags/intervals, as a reverse process of the above.

◆ ◆ ◆

72

GANITA SUTRA-11 AND GANITA SUTRA-12

1. Text of Ganita Sutra-11:

व्यष्टिसमिष्टिः *Vyasti-samastih*
Specific and General

2. Sequential organization of the text of Ganita Sutra-11:

01	02	03	04	05	06	07	08	09	10
व्	य्	अ	ष्	ट्	इ	स्	अ	म्	अ
11	12	13	14						
ष्	ट्	इ	ः						

Total Letters	Vowels	Nasels	Consonents
14	5	1	8

3. Text of Ganita Sutra-12:

शेषाण्यङ्केन चरमेण Sesnyankena Caramena
The Remainder by the last digit

4. Sequential organization of the text of Ganita Sutra-12:

01	02	03	04	05	06	07	08	09	10
श्	ए	ष्	आ	ण्	य्	अ	ङ्	क्	ए
11	12	13	14	15	16	17	18	19	20
न्	अ	च्	अ	र्	अ	म्	ए	ण्	अ

Total Letters	Vowels	Nasels	Consonents
20	9	-	11

5. Recapitulation:

The split of closed interval into closed interval and half closed interval, as such brings into focus a bigger/general format of closed interval as well as of smaller/specific closed interval. Likewise the split of a square with full boundary would permit a split into a square of full boundary and three others squares of not full boundary. Here as well there would be a bigger/general square format as well as of a smaller/specific square. Ahead as well there would be a bigger/general cube format and of smaller/specific cube. And so on there would be general and specific format hyper cubes.

6. Infinite range of closed interval:

Interval as a track of a moving point shall be bringing to focus a chase for closed interval in terms of infinite moving steps placements of points. This as such would be making a infinite steps mapping for every closed interval, may it be general or specific closed interval. Likewise would be the position for mappings of squares may it be of general or specific format as here square would be a track of infinite steps moments/placements of closed intervals in every such case. And ahead as well would be the parallel mapping situations for cubes as well as for hyper cubes.

This as such shall be bringing to focus the infinite steps chase of hyper cube (n) in terms of hyper cube (n – 1).

With it the deficiency (of the range of second parts like that of incomplete domains) which have been taken care of by Ganita Sutra-10 shall be making it the first part (as of complete domains) to be of features as the working rule of Ganita Sutra-11.

7. **Two fold enlightenment:**

   The fold ancient enlightenment : 'as in body so in universe', and 'as in universe so in body', is of the format of Ganita Sutra-11 working rule. It as such brings to focus the sequence of infinite number of replicas of general formats (of hyper cubes) as specific formats thereof. With it the transportation of the structural set-ups as packages becomes the feasibility feature of the systems.

8. **Place value organization format:**

   Taking illustrative case of our ten place value system as organization format for whole range of whole numbers by having coordination for the placements zones of values/whole numbers in terms of a sequence of values $10^0 = 01$, $10^1 = 10$, $10^2 = 100$ and so on, it can be noticed that the last digit of whole number organized in ten place value system, shall be having a value/digit '0' or '1 to 9' as of values of multiple '1'.

   However, along all other places the values would be of multiples '10'. As such the division by '10' shall be making available the value at the unit place/at the first place of value $10^0 = 1$ as the remainder (being the last digit).

   In case of division by 10^2 the number of values of first and second place that is at 10^0 and 10^1 as a double digit number shall be the remainder. Likewise would follow the remainder on divisions by 10^3 , 10^4 or in general by 10^N.

9. **Working rule of Ganita Sutra-12:**

   The working rule of Ganita Sutra-12 'शेषाण्यङ्.केन

चरमेण'/'Sesnyankena Caramena' (The Remainder by the last digit), as such is of the above division format of place value systems (may it be a ten place value system, or of any place value system).

10. Sankhiya Nishtha and Yoga Nishtha:

The above format of Ganita Sutra-12 availing features of place value system organization for artifices of numbers is as per the *Sankhiya Nishtha* approach. However the focus of the working rule of Ganita Sutra-10 'यावदूनम्'/'Yavadunam' (By deficiency (of double of it) avails the geometric format of 'incomplete domains' parts available on split of a complete domain into a complete domain part and incomplete domain part.

Here one may have a pause and permit the transcending mind to glimpse *Yoga Nishtha* ultimately availing artifices of numbers by presuming their geometric formats, and on the other hand *Yoga Nishtha* availing geometric formats presuming the existence of artifices of numbers.

Further, it would be a blissful exercise to permit the transcending mind to be face to face with the phenomenon of *Sankhiya Nishtha and Yoga Nishtha* being complementary and supplementary of each other and that the same at all steps running parallel to each other.

11. Parallel features of Sankhiya Nishtha and Yoga Nishtha:

It shall be a blissful exercise for the transcending mind to a remain in prolonged sittings of deep trans to be face to face with the phenomenon of artifices of numbers and of dimensional frames running parallel to each other. It is with this glimpsing that one shall be perfecting one's intelligence to be perfect with consciousness core as the source reservoir.

12. Transition from macro state to micro state:

The Transition from macro state domain to micro state domain shall be a transition from domain fold to dimension fold. It is the unison adhesive/glue which shall be structuring the dimensional domain. Its illustrative mathematics works as follows.

Step 1: 1-space in the role of dimension of 3-space.

Step 2: Pair of dimensions here (1-space as dimension) shall be requiring one unit for adhesive glue to be provided by dimension of 1-space i.e. by (–1) space.

Step 3: The mathematical equation.

(+1) because of first dimension +(+1) because of second dimensions (–1) (–1) because of adhesive glue equal to three units.

Step 4: A three dimensional domain would get structured out in the process.

The general mathematical equation for n-space in the role of dimension shall be : $(+ n) + (+ n) - (n - 2) = n + 2$, i.e. in the process would stand structured out (n + 2) space domain.

13. Structure because of three dimensions simultaneously under play:

To unite, third dimension to a set-up of pair of dimensions, shall be requiring a pair of adhesive/glue unit (of dimension of dimension) value, and it would work out as follows:

Step 1: Structural value because of pair of dimensions, as above shall be : n + 2.

Step 2: Now ahead, when third dimension would unite with the above structure of pair of dimension of value (n + 2) it shall be adding n because of third dimension but at the same time shall be requiring two glue units of value (n –2), which shall be, as such be the value (2 (n – 2) accordingly as a net result the mathematics would work out :

(n + 2) + n – (2) n – 2

6

As this is the general feature of the domain structured with the help of three dimensions of any dimensional order, as such the attainment would be of the order of space domain as an end result.

14. 6-space domain as an end result:

It may be recapitulated as that 4-space is the creator space (4-space). 4-space is the dimensional order of 6-space. Now let us chase the structural set-up because of three dimensions of the values of 4-space each.

Of these three dimensions of 4-space each, a pair out of them shall be constituting a 6-space structural set-up.

Further as that as far for every dimension, there would be two choices in terms of remaining two dimensions for having pairing with the startwith dimension and as such six structural set-up would be attainable twice.

Going like that, Ganita Sutra-1 to 6 followed by Ganita Sutras 7 to 12 shall be covered.

And, the remaining four Sutras shall be making it possible to set the structuring process in the reverse chase starting from Ganita Sutra-16 to Ganita Sutra-11, and from Ganita Sutra-10 to Ganita Sutra-5, and thereby reaching at the balance range of Ganita Sutras 4 to 1, to again have a reversal of the process, and thereby having a cyclic order for the organization format as well as of systems of Ganita Sutras.

◆ ◆ ◆

73

GANITA SUTRA-13 TO 16

1. Text of Ganita Sutra-13:

सोपान्त्यव्दमन्त्थम् Sopantyadvyamantyam
Completeness is at both ends

Text of Ganita Sutra-14

एकन्यूनेन पूर्वेण Ekanyunena Purvena
One less than Before

Text of Ganita Sutra-15

गुणितसमुच्चयः Gunitasamuccayah
The product of the Sum

Text of Ganita Sutra-16

गुणकसमुच्चयः Gunaksamuccayah
All the Multipliers

2. Transition from Ganita Sutra 12 to Ganita Sutra-13:

It may be recapitulated that sequential chase as of steps being Ganita Sutras-1 to 6 and Ganita Sutras-7 to 12 as such make the remainder range of Ganita Sutras-13 to 16 being a class in itself, as a third step in the above sequence and order of the chase.

Simultaneously as that in the reverse order Ganita Sutras-16 to 11 as well shall be making a single integrated step for the chase as such in the process, Ganita Sutras-11 and 12 because of its distinctive dual role, would deserve focused attention. It shall be of the feature of transiting from Ganita Sutra-1 to Ganita Sutra-12, as well as from Ganita Sutra-12 back to Ganita Sutra-11.

With it the sequential chase for reaching back from Ganita Sutra-12 to Ganita Sutra-11 a prior step features of reaching from Ganita Sutra-13 to Ganita Sutra-12 shall be having a special role to play.

Ganita Sutra-12 as such, under the place value system format, shall be having a focused attention upon the last unit value place of the format as remainder. This focused attention upon the last digit, as a step ahead shall be leading to focused attention upon the last two digits, of place values respectively being of unit and ten units values placed and to be processed simultaneously.

This pair of values namely unit and ten units, as such shall be making it a reflection pair (01, 10). It is this feature of this reflection pair which would help transcend from artifices format to dimensional spaces format.

One may have a pause here and permit the transcending mind to have a fresh look at the set-up of interval/ representative regular body of a regular body of 1-space. With 1-space content as domain fold of the four fold manifestation layer within creator space (4-space) (of spatial order) there shall be manifested a pair of end points for the interval making it a closed interval.

It is in terms of the manifested pair of end points for the interval making it a closed interval, as representative regular body of 1-space into a manifestation layer, shall be admitting (–1) space in the role of dimension of (+1) space. This as such shall be surfacing a pair of orientations for a line as going from one end to the second end and vice versa.

3. **Working rule of Ganita Sutra-13:**

   Let one have a pause here and have a glimpse of the working rule of Ganita Sutra-13 'सोपान्त्यद्वयमन्त्यम्'/'Sop antyadvyaman-tyam' (Completeness is at both ends). The focus here is upon the completeness at both ends. It is like having a 'source reservoir with a double feeder flow streams'.

One shall sit comfortably and permit the transcending mind to remain in prolonged sittings of deep trans to glimpse and to be face to face with the phenomenon of the interval (domain) as a tracking path of a moving point (for each end point) making the interval (domain) range as a feeder flow stream of the source reservoir. One may further have a pause here and permit the transcending mind to fully comprehend and to have a complete insight of the pair of feeder flow streams being of the linear formats as of opposite orientations. There being opposite orientations for the feeder flow streams linear formats, as such these shall be having a meeting point at the middle/centre/origin/source reservoir core.

In the context it would be relevant to note that the reflection pair (01, 10) as of placements at the pair of end points of a closed interval, shall be surfacing artifices values of different features of an interval. The difference value 10 – 01 = 09 = 9 = 4 + 1 + 4 shall be focusing upon the artifice value '5' as of the middle point placement value.

Here It would be relevant to note that hyper cube 5 accepts creative boundary of ten hyper cubes 4. As such the creative boundary shall be of 10×4 (40) spatial dimensional components/

coordinates. NVF (LINE) = 40 and NVF (INTERVAL) = 101; with 0 + 0 = 0 and pairing split as 10, 01 taken together shall be giving us an insight about 'line' and 'interval' and further as that 010 with $1 \times 1=1$ and the pairing split for 010 as (01, 10) and that NVF (STRAIGHT) = 102 = NVF (TWO SPACE) = NVF (A INTERVAL) =1 + 101 shall be further adding values 2 comprehension about the feature of 'STRAIGHT', and the way moving 'straight' line structures a square 'surface'.

4. **Working rule of Ganita Sutra-14:**

   The Working rule of Ganita Sutra-14 'एकन्यूनेन पूर्वेण'/ 'Ekanyunena Purvena' (One less than Before) focuses upon the second orientation of an interval, namely from second end to the first end track of a moving point of the reverse counting steps.

5. **Working rule of Ganita Sutra-15:**

   The Working rule of Ganita Sutra-15 'गुणितसमुच्चयः'/ 'Gunitasamuccayah' (The product of the Sum) focuses upon the simultaneous play of both the orientations as distinct axes like the pair of banks of the stream permitting constitution of the bed of the stream for flow currents of the stream.

6. **Working rule of Ganita Sutra-16:**

   The working rule of Ganita Sutra-16 'गुणकसमुच्चयः'/'Gunaksamuccayah' (All the Multipliers) focuses upon the whole range of flow currents at different layers of the stream bed and simultaneously unifying bed, stream bed flow layers. It would be like unified solids transiting and transforming into liquefied solids.

With it would be emerging through these sixteen Sutras range organization format being of the features of attainment of transition and transformation from first 'earth element' to that of second 'water element' which as a *Sankhiya Nishtha* shall be the attainment of transition and transformation from artifice-1 to artifice 2 and as *Yoga Nishtha* being the attainment of transition and transformation from linear order 3-space to spatial order 4-space.

As such as a reverse process from Ganita Sutra-16 to Ganita Sutra-1 it in a continuity shall be attaining transition from second element, namely water element to third element i.e. 'fire element' and the cyclic process again shall be sequentially taking fire element, air element and back from air element to space element.

A step ahead shall be a reach from space element to Sun, and ahead from Sun to Pole Star. Still further it shall be taking from Pole star to nature (and from nature to Braham) and further ahead from Braham to Par Braham.

7. **Bliss for Minds fulfilled with intensified urge:**

   Bliss of values and virtues flourishes in full bloom for the sadhkas whose hearts and minds are fulfilled with intensity of urge for the enlightenment to continuously remain on Ganita Sutras format for sequential chase starting with the first element (Earth and attaining transitions and transformations step by step to be on the feeders line of the source reservoir and to acquire human enlightenment and being worthy recipient of *Brahaman* grace to be in unison with Par Braham.

◆ ◆ ◆

74

OVERVIEW OF VEDAS REACHING US

1. Vedic wisdom reaching us transcending through Times is of two broad reservoirs, firstly as the Vedic traditions reservoir and secondly as the Written Word. Both are complementary and supplementary of each other. Each helps very much to under stand the other.
2. The Written Word (of vedic wisdom) centres around *Samhita*, of them as well the focal placement is for the *Sakala Rigved Samhita*.
3. It is blissful for us that the *Sakala Rigved Samhita* has reached us fully intact from its first letter (syllable) to the last syllable (letter) of the entire range of the *Samhita*. The text range of *Sakala Rigved Samhita* in its written form is of the range of 397265 syllables of its knowledge range of 432000 syllables. The remaining 34735 syllables constitute a un manifest text which enliven itself as *Balya Khillay* mantras.
4. Tradition goes as that every other *Samhita* of Branches of *Rigved* as well as of other Vedas can be enlivened with the Vedic knowledge system of *Rigved Samhita* itself.
5. With it we can feel privileged to be in possession of

the whole range of Vedic systems, knowledge and wisdom.

6. Vedic traditions reaching us make us conscious as that the range of Written Word which should have been available with us is to be as of all the *samhitas* of 21 branches of *Rigved*, 101 branches of *Yajurved* and of 1000 branches of *Samved* and 9 branches of *Atharav Ved*. However of these 1131 Samhitas only, say ten of them have reached us intact and the remaining could not transcend the time and to decend upon us. As such it is going to be, really a very big reconstruction work assignment being their remaining to be attended by the present and the future generation.
7. The other available Vedic literature inspired by Vedic enlightenment, as such may be availed by the Sadhkas fulfilled with intensity of urge to attend to the unfinished work of enlivening and reconstructing the written Vedic Word which as not reached us.

◆ ◆ ◆

75

GANITA UPSUTRAS

1. Ganita Sutras and Upsutras:

(*a*) Ganita Sutras and Upsutras are complementary and supplementary and these together constitute a single integrated system.

(*b*) There are sixteen Sutras and thirteen Upsutras.

(*c*) Ganita Sutras as well as Ganita Upsutras individually as well are self sustained systems as well.

(*d*) There are many ways to approach Ganita Upsutras as integral part of the integrated system of Ganita Sutras and Upsutras.

(*e*) One way to approach Ganita Sutras is an independent system.

(*f*) Ganita Upsutras, as an independent system, may be approached Upsutra wise, in the sequence and order of their placement as Ganita Upsutra-1 to Ganita Upsutra-13.

(*g*) Simultaneously references may be had to the relevant Ganita Sutras to which the consent Upsutras have more affinity.

(*h*) It is like Ganita Upsutra-1 may be taken us having more affinity with Ganita Sutra-1 as well as with Ganita Sutra-2. In a way it may be taken as that, as in integrated system the sequence would follow as 'Ganita Sutra-1', Ganita Upsutra-1 and Ganita Upsutra-2

(*i*) This way, it may be taken that Ganita Sutra-1 to 13 may be approached sequentially as steps subsequent to Ganita Sutra-1 to 13, as gaps filling processes of Ganita Sutras.

(*j*) In this process, the last three Ganita Sutras may emerge as of features permitting no gaps in between.

2. Table of the working rules of Ganita Sutras comes to be as follows:

(S01) By One More than One before

(S02) All From 9 and the Last from ten

(S03) Vertically and Crosswise

(S04) Transpose and Apply

(S05) If the Samuccaya is the same it is Zero

(S06) If one is in Ration the others is Zero

(S07) By Addition and by Subtraction

(S08) By the Completion or Non-completion

(S09) Differentiation Calculus

(S10) By the Deficiency

(S11) Specific and General

(S12) The Remainder by the Last Digit

(S13) The Ultimate and twice the Penultimate

(S14) By One less than the One Before

(S15) The Product of the Sum

(S16) All the Multipliers

3. Text of Ganita Upsutras:

Here is being reproduced text of Ganita Upsutras, in original as well as in *Devnagri* script along with simple English Rendering for their working rules.

(01) आनुरूप्येण (Anurupyena)/Proportionately

(02) शिष्यते शेषसंज्ञ (*Sisyate Sesasamjnah*)/That remains is remainder

(03) आद्यमाद्येन अन्त्यमन्त्येन (*Adyamadyen Antyamantyena*)/ First with first Last with last

(04) केवलैः सप्तकं गुण्यात् (*Kevalaih Saptakam Gunyat*)/Only Seven as multiplicand

(05) वेष्टनम (*Vestanam*)/Osculators

(06) यावदूनं तावदूनम (*Yavadunam Tavadunam*/ That twice This twice

(07) यावदूनम् तावदूनीकृत्य वर्ग च योजयेत् (*Yavadunam Tavaduni-krtya Varganca Yogayet*)/That twice This twice Square and add

(08) अन्त्पयोर्दषकेऽपि (*Antyayordasake'pi*)/Ends to sum as ten

(09) अन्तययोरेव (*Antyayoreva*)/Ends to be in ratio

(10) समुच्चयगुणितः (*Samuccayagunitah*)/Samuchya as product

(11) लोपन स्थापनाभ्याम् (*Lopana Sthapananabhyam*)/That missing to be established

(12) विलोकनम् (*Vilokanam*)/By observation

(13) गुणितसमुच्चयः समुच्चयगुणितः (*GunitaSamuccaya Samucc-ayagunitah*)/Product samuchya Samuchya Product

4. The table of working rules in their English rendering expressions comes to be as follows:

(U01) Proportionately

(U02) That remains is remainder

(U03) First with first Last with last
(U04) Only Seven as Multiplicand
(U05) Osculators
(U06) That twice This twice
(U07) That twice This twice Square and add
(U08) Ends to sum as Ten
(U09) Ends to be in Ratio
(U10) Samuchya as Product
(U11) That Missing to be Established
(U12) By Observation
(U13) Product *samuchya Samuchya* Product.

5. Integrated placement for working rules Ganita Sutras along with Ganita Upsutras in their combined sequential formats comes to be as follows:

(S01) By One More than One before
(U01) Proportionately
(S02) All from 9 and the Last from Ten
(U02) That Remains is Remainder
(S03) Vertically and Crosswise
(U03) First with first Last with last
(S04) Transpose and Apply
(U04) Only Seven as Multiplicand
(S05) If the Samuccaya is the same it is Zero
(U05) Osculators
(S06) If one is in Ration the others is Zero
(U06) That twice This twice
(S07) By Addition and by Subtraction
(U07) That twice This twice Square and add
(S08) By the Completion or Non-completion
(U08) Ends to Sum as Ten
(S09) Differentiation Calculus

(U09) Ends to be in Ratio
(S10) By the Deficiency
(U10) Samuchya as Product
(S11) Specific and General
(U11) That Missing to be Established
(S12) The Remainder by the Last Digit
(U12) By Observation
(S13) The Ultimate and Twice the Penultimate
(U13) Product *Samuchya* *Samuchya* Product
(S14) By One less than the One Before
(S15) The Product of the Sum
(S16) All the Multipliers

6. There are two basic established processing approaches, namely **Sankhiya Nishtha and Yoga Nishtha.** Of these the *Sankhiya Nishtha* avails artifices of numbers, while the *Yoga Nishtha* avails the geometric formats. Both these approaches are self-sustained approaches and run parallel to each other. However these are complementary and supplementary of each other.

7. Here, the **Sankhiya Nishtha** approach is being taken up for pointed attention as to the working rules of Ganita Upsutras with the help of illustrative applications thereof.

◆ ◆ ◆

76

VEDIC MATHEMATICS APPROACH

1. Pure and applied values:

(*i*) Mathematics has pure and applied values.

(*ii*) Pure and applied values of mathematics can be approached distinctively as well as simultaneously.

(*iii*) Of it, the Vedic mathematics approach weighs heavily in favour of pure and applied values to be reached at simultaneous.

(*iv*) It is there as the vedic systems acceptance of *Sankhiya Nishtha and Yoga Nishtha* being complemen-tary and supplementary of each other, as well as that these run parallel to each other.

2. Sankhiya Nishtha and Yoga Nishtha:

(*i*) *Sankhiya Nishtha* presume the existence of geometric formats and avails artifices of numbers to work out the entire existence phenomenon and whole range of knowledge with mathematics at its core.

(*ii*) *Yoga Nishtha* presumes the existence of artifices of numbers and avails dimensional frames for chase of existence phenomenon in its entirety as of geometric formats.

(*iii*) With it it becomes axiomatic as that artifices of numbers and dimensional frames run parallel to each other.

(*iv*) Accordingly the number systems and geometric formats become complementary and supplementary of Vedic mathematics systems.

3. Arithmetic, Geometry and Algebra:

(*i*) Vedic mathematics approach, as such becomes of three folds, parallel to the values of artifices of numbers dimensional frames and interrelationship of artifices of numbers and geometric formats.

(*ii*) Arithmetic predominantly is about features and values of numbers and artifices of numbers.

(*iii*) Geometry pre-dominantly is about the features and values of bodies and dimensional frames.

(*iv*) Algebra is predominantly about the features and values of interrelationship artifices of numbers and dimensional frames.

4. Arithmetic system:

(*i*) Arithmetic system may as well be designated and taken as numbers system.

(*ii*) Whole numbers together with four basic (arithmetic) operations, namely (addition, subtraction, multiplication and division) is start with set of features and values of numbers.

(*iii*) A step ahead would emerge rational numbers, irrational numbers, transcendental numbers, self-referral numbers and unity state numbers and correspondingly would get extended the range of operations in terms of which different features and values of these sets of numbers can be reached at. The corresponding geometric formats would help have full insight for having complete comprehensions of features and values of these sets of numbers and their distinctive operational systems.

5. **Geometric systems:**
   (*i*) Geometric systems essentially are dimensional centric.
   (*ii*) Basically, it is the dimensional order of a space which leads to the distinctive features and values of the concerned dimensional domains.
   (*iii*) Vedic systems primarily work out a group of six real spaces, designated and known as 1-space, 2-space, 3-space, 4-space and 5-space with interval, square, cube and hyper cubes 4, 5, 6 being their respective representative regular bodies.
   (*iv*) With it point, line, surface, solid and hyper solids 4, 5, 6 become the basic sets of features and values of spaces 0, 1, 2, 3, 4, 5 and 6.
6. **Algebraic system:**
   (*i*) Basic role of algebraic to work out the features and values of interrelationships of arithmetic systems and geometric systems.
   (*ii*) Algebra accept geometric formats/bodies/spaces as sets of structured points as elements permitting given operations (of arithmetic values as well as of geometric values).
   (*iii*) To reach arithmetic values for the given geometric formats/bodies/domains within dimensional frames is one of the aims of algebraic system.
   (*iv*) Further to reach at geometric values for the set of features and values of artifices of numbers is the other pre-dominant aim of algebraic system.

7. Unified states:

(*i*) Vedic systems as such, as a result of complementary and supplementary approach of *Sankhiya Nishtha and Yoga Nishtha* take us to the unified states of existence phenomenon in its entirety including that of existence phenomenon within human frame.

(*ii*) Students of Vedic mathematics as such as a result of their comprehension and imbibing of the features and values of Vedic mathematics systems come face to face with their range of consciousness states.

(*iii*) This ultimately helps attain unity state of consciousness.

It is this attainment which is going to be the bliss and fruit of the Vedic mathematics approach.

◆ ◆ ◆

77

VEDIC MATHEMATICS DIRECTION

1. Mathematical mind:

(*i*) Mathematics, as a Discipline, is to be definite about the ultimate 'mathematical mind' it institutionalizes.

(*ii*) No doubt it is the mind which creates and stabilizes the discipline of mathematics and in the process institutionalizes itself as well as a 'mathematical mind' of the order of the mathematical domain created by the mind itself.

(*iii*) In the process of creation and having formatted itself of that order and features makes it as to be of the 'values and virtues' of the emerging 'intelligence field'.

(*iv*) It is this transition and transformation for the mind's creation into its own formatting as 'intelligence field', which deserves to be chased for 'modern mathematical mind' vis-à-vis ancient wisdom preserved with us.

2. Modern mathematical mind:

(*i*) From the mathematics problem at which the modern mathematical mind find struck itself from sixteen century till date, it leads to the conclusion that the central cause for it is in the modern mathematical approach as to be of 'linear order thinking'.

(*ii*) The other cause which comes to focus is the mis-directed idealism of abstraction by taking as if, while firstly mind can be isolated from as if the same is not part of the existence phenomenon, and secondly as if even mind as well can be dispensed with in the mathematical theories, all in the name of 'objectivity', rather a newo mathematical objectivity.

(*iii*) Lines to curves to spirals to parallel direct and reverse orientations, and all that in that sequence are attempts of the modern mathematics to dispense with the linear order and to be fully free of it but this all is not becoming possible for it and still the insistence to continue pursuing it like that is creating all knots and the modern mathematics problem.

(*iv*) Amongst others, the cut-off from the 'past' approach in the name of an independent approach is proving like attaining continuity by jumps as is happening in the attempt to encircle surfaces and while reaching at the centreed point taking it to be a cipher or zero but remaining void of mathematics of zero space in the role of dimensional order of 2-space.

3. Vedic mathematical direction:

(*i*) Modern mind needs Vedic mathematics directions for having transition and transformation for its linear thinking process into that of sequential multi dimensional orders processing systems.

(*ii*) Primes and space filling curves like absorptions

are not providing needed transitions and transformations for linear thinking processes, and for it, the modern mind on following Vedic mathematical direction may attain desired breakthrough for its systems beginning with the start with transition from linear order to that of spatial order, and ahead from spatial order to solid order, and from solid order to hyper solid orders.

(*iii*) This as such shall be firstly providing a take off from mathematics of '1' to that of mathematics '2' and ahead from mathematics of '2' mathematics of '3' and so on.

(*iv*) The classical and orthodox type problems of mathematics with which the modern mind is struck at are only because of mathematics of '1', particularly when it is attempted to be worked out with half as one and sticking to linear order/line as a format and unknowingly the mathematics so worked leads to embracing situations of the function being every where continuous but being no where derivable.

4. Spatial order:

(*i*) Spatial order, as comparison to linear order (which yields mathematics of one) it yields mathematics of two.

(*ii*) In other words linear order, which yields mathematics of one, accommodate numbers along a line, to be designated as a number line. However spatial order, which yields mathematics of '2' accommodates the numbers as a surface/plane area; illustratively, '1' becomes '1×1'; '2' becomes '1×2' and so on '3' as '1 3', '4' as '1×4'.

(*iii*) As linear order works out space in terms of linear dimensions (lines as axes/1-space in the role of dimension), the spatial order works out the space in terms of spatial dimensions (2-space in the role of dimension).

(*iv*) The linear order structure the space as 3-space (three linear dimensions space) and with it the whole chase within three dimensional frame (of three linear dimensions), as three space becomes linear in order and as a result the whole thinking process as well as of the attainments domains stands restricted to 'linear order', as is happening with the modern mathematics and its system as well as with the formatting of modern mathematical mind. However, on the other hand, the spatial order shall be availing 2-space in the role of dimension which would structure the space around as a 4-space confined within dimensional frames of four spatial dimensions.

(*v*) Vedic system accept it a creator space (4-space).

5. Transition from 3-space to 4-space:

(*i*) By following Vedic mathematics direction of shifting from linear dimensional order (1-space in the role of dimension which structures the space around as 3-space within three dimensional frame of three linear dimensions) to spatial dimensional order which shall be structuring space around as 4-space of four spatial dimensions, the whole range of features, values and virtues of linear order mathematics would automatically get transformed into the new set of features, values and virtues.

(*ii*) The interval as part of line and as such 1-space body and it its role as linear dimension/axis of 3-space, shall be transiting and transforming acquiring new set of features, values and virtues of spatial order 4-space by 'interval' becoming 'hyper cube-1' formats of a four fold manifestation layer of quadruple values (–1, 0, 1, 2)/(–1-space in the role of dimension of 1-space, 0-space in the role of boundary of 1-space), 1-space itself in the role of domain fold of 1-space and 2-space being in the role of origin of 1-space.

(*iii*) This as such, shall be bringing us face to face with the existing approach of modern mathematics where even to pause a question as to what is the dimensional order of 1-space, may be dreaded as same being unthinkable so far, and this as such, shall be intensifying the urge to follow the Vedic mathematics directions as here only lies the hope for the linear order based modern mathematical thinking processes to have required liberation from the confinements of 'linear order 3-space'.

◆ ◆ ◆

78

VEDIC MATHEMATICS (Ganita Sutras Steps for Enlightenment)

Introductory

1. Ganita Sutras is a complete scripture.
2. Ganita Sutras, being a complete scripture, unfold from within itself.
3. Ganita Sutras are to be understood in terms of Ganita Sutras themselves for both pure knowledge as well as for organization of knowledge.
4. This as such would mean that the Ganita Sutras text is to be approached in the sequence and order of its composition.
5. Each Sutra (and Upsutra) is complete in itself but they maintain the sequence and order of their placements in the text of Sutras (as well as in the text of Upsutras).
6. Sutras are followed by Upsutras, and looking from other way round, it may be taken as that Sutras are followed by Sutras.
7. Sutras avail the artifice 16, which is parallel to the values and virtues of '*Shodas Kalan* (16 values skills)'

and Ganita Upsutras avail artifice 13 parallel to the values of '*Dhan Teras* (13 enrichment features).

8. Artifice 16 of 10 place value, when is taken as 7 place value, it becomes equivalent to artifice 13 of 10 place value as that 16 = 10 + 6 and 13 = 7 + 6.
9. The text of Ganita Sutras (including Ganita Upsutras) avails in all 520 letters, including the transcendental source reservoir sole syllable Om (ॐ).
10. This, this way makes a long range of steps, precisely 520 steps are to be followed to complete its chase.
11. Taking availability of transcendental source reservoir in the form of sole syllable Om (ॐ), the first chase step, in the background would emerge as to be of the features, order, values and virtues of the first letter of the text of Ganita Sutra-1, with which the enlightenment chase for the sadhkas/seekers fulfilled with intensity of urge to know and to have enlightenment, is to begin.

First Step

11. First letter of Ganita Sutra-1 'एकाधिकेन पूर्वेण'/(*Ekadhiken Purvena*) is 'ए', the sixth vowel.
12. The first letter, namely 'ए' being the sixth vowel, it as such accepts association of artifice '6'.
13. *Sankhiya Nishtha* which on ultimate analysis avails artifices of numbers, presume the existence of dimensional frames.
14. This, this way would bring into play, parallel to artifice 6, a six dimensional frame, and as such a 6-space, as well as hyper cube 6, being the representative regular body of 6-space.
15. Artifice 6 as number 6, is the number of 'Sun'.

16. Sun as atman/sole, and Sun as Vishnu lok, that way, simultaneously bring into the values and virtues of Sun, Atman and Vishnu lok.
17. Lord Vishnu as the presiding deity of a measuring rod (the *sathapatya* measuring rod) as such makes available the sequential approach path of the measuring rod of 6-space, which sequentially is to take us from representative regular bodies of 1-space to 6-space itself, namely, interval, square, cube and hyper cubes 4, 5 and 6.
18. As this measuring rod of these 6 sequential steps is of the features of Shad Chakra format of human frame, as such the Pursha format as well comes into play at this very first step of enlightenment of Ganita Sutras.
19. This as a result makes this first step of enlightenment as to be of the features, order, values and virtues of (1) artifice 6 (2) 6-space (3) hyper cube 6 (4) Sun (5) Lord Vishnu and (6) *Pursha format*.
20. From these 6 formats point of view, there can be 6 different ways of imitations for enlightenment in terms of the Ganita Sutras.
21. While remaining focused to artifices of numbers, the startwith focus naturally would be upon 'artifice 6'.
22. With availability of the transcendental source reservoir of sole syllable Om (ॐ), the Brahman values/Braham virtues, which in the context of *Sankhiya Nishtha* processing approach shall be the '*Nav Braham* (नव ब्रह्म)' as nav (नव) means new as well as nine as such the source reservoir value for artifice 6, in the circumstances is to be taken as 'artifice 9'.
23. This, this way shall be making the starting point as to be of source reservoir features for artifice '6'.
24. The artifices pair (9, 6) would become the source artifices

pair for reaching at the features, order, values and virtues of 'Sun' vis-à-vis 'Braham'.

25. The second letter of the text of Ganita Sutra-1 'एकाधिकेन पूर्वेण' being 'क' being the first varga consonant, having the meaning, values and virtues of 'Brahma', the four head Lord, creator the Supreme, the presiding deity of 4-space, as such the pair of artifices (9, 6) shall be leading to triple artifices (9, 6, 4).
26. One of the features of this triple artifices (9, 6, 4) is that $9 = 6 \times 1\frac{1}{2} = 4 \times 1\frac{1}{2} \times 1\frac{1}{2}$.
27. The artifice 6, the middle artifice of the triple (9, 6, 4), as such acquires distinctive features, as that $1 + 2 + 3 = 1 \times 2 \times 3 = 2 + 2 + 2$.
28. This feature as a transition feature for the artifices pair (9, 6) comes to be (3 + 3 + 3, 2 + 2 + 2), a step ahead the artifices pair (6, 4) is of the features (3 + 3, 2 + 2).
29. As such the triple artifices (9, 6, 4) leads to a sequential transition steps as (*i*) (3 + 3 + 3, 2 + 2 + 2) (*ii*) (3 + 3, 2 + 2).
30. This in a sequence shall be leading to (*i*) (3 + 3 + 3, 2 + 2 + 2) (*ii*) (3 + 3, 2 + 2) and (*iii*) (3, 2).
31. It would be relevant to note that 3 + 2 = 5/5-space/ transcendental (5-space) domain presided by Lord Shiv, the five head Lord with three eyes in each head, shall be leading to the features $3 \times 5 = 15$ and that the first *varga* consonant, namely 'क', as well is acquiring the feature, order, values and virtues of Lord Shiv.
32. It is this simultaneous features, orders, values and virtues for the first *varga* consonant (क) as Lord Brahma, as well as Lord Shiv shall be leading to enlightenment of the way spatial order 4-space set-up permits transcendence into solid order 5-space.
33. It is this phenomenon which need be completely comprehended and to be fully imbibed for initiations

for the Vedic systems being organized as sequential steps as text of Ganita Sutras.

34. This first step of enlightenment while accepts Braham as the source reservoir and with this source reservoir wants to reach at the features, order, values and virtues of 'Sun'.
35. With it Sun God, becomes the god of the first enlightenment step of the Ganita Sutras.
36. From this step, it is the enlightenment of the Sun as its features, orders, values and virtues which have to lead us to the second enlightenment step as of features, orders, values and virtues of the second letter of the text of Ganita Sutra-1, namely the first varga consonant letter 'd' as Lord Brahma as well as Lord Shiv are the presiding deities of this phase and stage of enlightenment which are personified as 'parents', and as such for enlightenment of this phase and stage of second step, 'parents' have to play the role of God for the innocent minds.
37. It is this shift from the 'Sun' to parents, which is the fruit of enlightenment of this enlightenment step.

Second Step

38. The second letter of Ganita Sutra-1 'एकाधिकेन पूर्वेण' is 'क' first *varga* consonant.
39. The transition from first step letter 'ए', the sixth vowel to the second step letter 'd' is the transition from the grace of Lord Vishnu, the overlord of 6-space to lord Brahma (and Lord Shiv), the presiding deity (s) of 4-space (and 5-space).
40. This infact would be a transition from Sun God to parents as God.
41. Sequentially, it also would mean a transition from Ganita Sutra-1 to Ganita Sutra-2.

42. Here it would be relevant to mention, and for the present to be taken as postulate as that 16 Ganita Sutras unfold parallel to 16 letters of the text of Ganita Sutra-1 itself.
43. It is the parallel sequence and order of 16 letters of Ganita Sutra-1 and sequential order of 16 Ganita Sutras, starting with Ganita Sutra-1 uptill Ganita Sutra-16, are to be taken as the structural keys of these Sutras.
44. First letter 'ए' of Ganita Sutra-1 as such is to supply the structural keys for Ganita Sutra-1.
45. A step ahead, second letter 'क' of Ganita Sutra-1 shall be supplying the structural key for Ganita Sutra-2.
46. With Sun god as the source reservoir of the features, order, values and virtues of Ganita Sutra-1 is to be looked for the comprehension of features, order values and virtues of Ganita Sutra-1.
47. It is by imbibing the features, order, values and virtues of 'Sun' as that one can have an enlightenment about the 'features, order, values and virtues of Ganita Sutra-1'.
48. The source reservoir for the features, order, values and virtues for Sun is Braham/sole syllable Om (ॐ).
49. The source reservoir for features, order, values and virtues is the 'second letter' of Ganita Sutra-1.
50. The innocence of the child is to be dependent upon the enlightenment of the parents.
51. The enlightenment of the parents is to follow the features, orders, values and virtues of Lord Brahma, creator the supreme.
52. This enlightenment is to be further intensified with the transcendence process within creator space (4-space) at its origin to the transcendental (5-space) domains.
53. Parents as 'God' are under secrete duty to play the role of god for the innocent young minds to prepare

them for the needed enlightenment of the creator space (4-space) with transcendental (5-space) domain as its source origin.

54. The innocent mind being of affine state it is to be properly formatted by the parents particularly the mother as to be parallel to the features, order, values and virtues.
55. For it, naturally the parents have to go to the shelter of Sun God for the Brahman enlightenment to be imparted to the innocent young minds by them.
56. The second letter, namely the first varga consonant 'क' firstly being of the format of creator space (4-space), which is of spatial order, and as such, every care is to be taken to ensure that the innocent young minds are not conditioned by the linear order for which the formulation '1' is to play.
57. One may have a pause here and to be ensure as that it is the biggest responsibility for the parents as God for innocent young minds that the innocent young minds are not in any way conditioned by the linear order.
58. Here, as such, the need would be to sure as that the young mind get smooth transition from artifice 1 to artifice 2.
59. For it the help can be had from the features, order, values and virtues of Ganita Uputra-1 'आनुरूप्येण (*Anurupyena*)'.
60. It is the symmetry as of the proportionate value, which is to be properly availed for this objective.
61. That being so, the focus of parents for proper formatting of innocent young mind, would remain centreed firstly upon the artifice 1.
62. The second fold of the focus is to centre around the artifice 2, and artifices ahead.

63. This focus is to be with proper caution that artifice 2 simply does not remain 1 + 1.
64. The focus as well is to be seriously upon the organization for artifice 2 as 1×2, in the background of artifice $1 = 2^0$.
65. To have proper appreciation, one shall pause to oneself as to what would be the next, namely 'third' artifice value ahead of the pair of values '1, 2'?
66. The first natural question would be that this artifice ahead would be '3'.
67. However, one may again have a pause here and think, meditate and transcend through the pair '1, 2' and to see if there can be any other answer, other than the answer '3'?.
68. One may again have a pause here and to reflect upon the re-organization for the pair (1, 2) as $(2^0, 2^1)$.
69. This re-organized set-up for the pair (1, 2) as $(2^0, 2^1)$ shall be helping have a pair of triples namely (1, 2, 3) and $(2^0, 2^1, 2^2)$.
70. Of the second triple $(2^0, 2^1, 2^2)$ the lead would be as that this triple is of values (1, 2, 4).
71. It is like this that artifices of numbers result into knots and make them a problem to crossover.
72. One may again have a pause here and to pause to oneself as to why the modern mind is frequently being confronted with 'mathematical problem'.
73. All these problem are there because of the axioms and also because of non sequential jumps being permitted to the mind to have a cross over the sequential steps and to have conclusions without verification and hence the postulates and problems.
74. It is this what is to be avoided by proper formatting of the innocent young mind in a most appropriate way at the earliest formatting phases and stages of the young minds.

75. It is here where the parents are under solemn obligation and duty towards innocent young minds.
76. One may again have a pause and permit oneself to have a comprehensive view of the existence phenomenon around, including that of human being existence within human frame, being designated as 'body'.
77. It would come to knowledge as that the created world around being of the four fold manifestation format, the domains here are sealed at the origin fold.
78. The origin fold being sealed, the transcendence there from, naturally is to be of appropriate steps.
79. As is evident from the two fold state of first *varga* consonant 'क' as of four fold creator space (4-space) and also as of solid order transcendental (5-space) domain, the compactification of this pair of states is like that of a surface standing compactifying with solids and the transcendence from surface to domain being permissible without much hindrance.
80. One may again have a pause here and permit the transcending mind to be face to face with the transcendence phenomenon of Sun light to the manifested creations around us including of our body whereby the inner fold micro state bodies become lively and vibrant in the process.
81. As such parents have to avail the grace of 'Sun' as 'Sun rays' reaching us.
82. As such the phenomenon of simultaneous existence of a pair of states need be comprehended and imbibed for their features, order, values and virtues for availing them for the timely formatting of the innocent young minds.
83. Here one may again have a pause and make oneself conscious as that, here slowly we are transiting from 'artifices of numbers' to 'geometric format of existence phenomenon'.

84. This infact is a transition from *Sankhiya Nishtha approach to Yoga Nishtha* approach.
85. Sankhiya Nishtha approach is of artifices of numbers of the geometric format.
86. On the other hand, *Yoga Nishtha* approach avails geometric formats along the artifices of numbers base.
87. One may again have a pause and to remind oneself as that Ganita Sutras are to permit six fold approach, as numerated above, as along (1) artifice 6 (2) 6-space (3) hyper cube 6 (4) Sun (5) Lord Vishnu and (6) *Pursha format*.
88. Broadly when *Sankhiya Nishtha* approach is being worked for unfolding Ganita Sutras, the same would lead to *Yoga Nishtha* approach for Ganita Upsutras.
89. Illustratively, the artifice 6 beginning for Ganita Sutras shall be leading to hyper cube 6 beginning for Ganita Upsutras.
90. However, it also would be permissible to go the other way around and to have hyper cube 6 beginning for Ganita Sutras and artifice 6 beginning for Upsutras.
91. This is being focused here as that, it would be solemn responsibility of the parents to see that the innocent young minds are properly formatted for *Sankhiya Nishtha and Yoga Nishtha* being complementary and supplementary processes running parallel to each other at every step.
92. This, this way while being viewed from the phase and stage of Ganita Sutra-2 back to Ganita Sutra-1 straight or through Ganita Upsutra-1, it shall be requiring re-visit to the formulation '1'.
93. Here it would be relevant to note that, the sixth vowel 'ऊ', naturally presume the existence of five previous vowel.

94. It would take us to second *Maheshwara Sutra* 'अक्' to comprehend as that the first five vowels 'अ, इ, उ, ऋ, ऌ' range is coordinated and is bounded by 'क्'.
95. This as such is a presumed existence prior to the formulation 'एक'.
96. It as such is parallel to the presumed existence of *Panch Mahabhut*/five basic elements, namely Earth, Water, Fire, Air and Space for the 'Sun'.
97. As such parents have to be extra conscious while formatting innocent young minds, as that the frequencies of Sun light are to be super imposed the frequencies of sole syllable Om (ॐ).
98. It is this paired set of frequencies, which shall be attaining the needful compactified and paired states for the phase and stage of the second God/Parents at the second step of enlightenment.
99. With focus upon artifices of numbers, the transition from the phase and stage of working rule of Ganita Sutra-1 as one more than before, shall be to the place value system of the rule 'all from nine and last from ten'.

Third Step

100. The third letter of Ganita Sutra-1 'एकाधिकेन पूर्वेण' is the elongated first vowel 'आ'.
101. The transition from second step letter 'क्', the first *varga* consonant to the third step letter 'आ', the elongated first vowel is the transition from the grace of 'parents' to the grace of *Acharya* 'आचार्य'/Fully cultured Teacher/ Guru.
102. With it the responsibility for initiations and instructions shifts from parents to the solemn responsibility of *Acharya* 'आचार्य'.

103. One may have a pause here and re-glimpse the sequential transition step from the transcendental source reservoir Brahman domain of sole syllable Om (ॐ) to the domain of Sun God to the domain of enlighte-ned parents to the domain of fully cultured teacher.

104. *Acharya* shall be explaining mathematical approach as well as shall be giving mathematical directions and simultaneously shall be settling Vedic mathematics learning steps, learning progression charts, sequential learning progression flow, in the light of the working rules of Ganita Sutras.

◆ ◆ ◆

Annexure-1

TEXT GANITA SUTRAS

Sutra No.	**Sutra text** (Original in Sanskrit, with Roman script and with simple English Rendering)
01.	ॐ । एकाधिकेन पूर्वेण । *Om* । *Ekadhiken Purvena.* Om । By One More than One before.
02.	निखिलं नवतश्चरमं दशतः । *Nikhilam Navatascramam Dasatah.* All from 9 and the last from ten.
03.	ऊर्ध्वतिर्यग्भ्याम् । *Urdhva tiryagbhyam.* Vertically and crosswise.
04.	परावर्त्य योजयेत् । *Paravartya Yojayet.* Transpose and Apply.
05.	शून्यं साम्समुच्चये । *Sunyam Samyasamuccaye.* If the samuccaya is the same it is Zero.
06.	(आनुरूप्ये) शून्मन्यत् । *(Anurupye) Sunyamanyat.* If one is in Ration the others is Zero.

07.	संकलव्यवकलनाभ्याम्। *Sankalana-vyavakalanbhyam.* By addition and by subtraction.
08.	पूरणापूरणाभ्याम्। *Puranapuranabhyan.* By the completion or non-completion.
09.	चलनकलनाभ्याम्। *Calana-kalanabhyam.* Differentiation Calculus.
10.	यावदूनम् *Yavadunam.* By the Deficiency.
11.	व्यष्टिसमिष्टि *Vyastisamastih.* Specific and General.
12.	शेषाण्यङ्केन चरमेण। *SesnyankenaCaramena.* The Remainder by the last digit.
13	सोपान्त्यव्दमन्त्यम्। *Sopantyadvyamantyam.* The ultimate and twice the penultimate.
14	एकन्यूनेन पूर्वेण। *Ekanyunena Purvena.* By One less than the One Before.
15	गुणितसमुच्चयः। *Gunitasamuccayah.* The product of the Sum.
16	गुणकसमुच्चयः। *Gunaksamuccayah.* All the Multipliers.

Annexure-2

TEXT GANITA UPSUTRAS

Upsutra No.	**Upsutra text** (Original in Sanskrit, with Roman script and with simple English Rendering)
01.	आनुरूप्ये । *Anurupyena.* Proportionately.
02.	शिष्यते शेषसज्ञः । *Sisyate Sesasamjnah.* That remains is remainder.
03.	आधमाधेनान्त्यमन्त्येन। *Adyamadyen Antyamantyen.* First with first Last with last
04.	केवलैः सप्तकं गुण्यात् । *Kevalaih Saptakam Gunyat.* Only Seven as multiplicand
05.	वेष्टनम् । *Vestanam.* Osculators.
06.	यावदूनं तावदूनम् । *Yavadunam Tavadunam.* That twice This twice.

07.	यावदूनं तावदूनीकृत्य वर्गं च योजयेत्। *YavadunamTavadunikrtyaVargancaYogayet.* That twice This twice Square and add.
08.	अन्त्ययोर्दशकेअपि। *Antyayordasake'pi.* Ends to sum as ten.
09.	अन्त्ययोरेव। *Antyayoreva.* Ends to be in ratio.
10.	समुच्चयगुणितः। *Samuccayagunitah.* Samuchya as product.
11.	लोपनस्थापनाभ्याम्। *Lopana Sthapananabhyam.* That missing to be established.
12.	विलोकनम्। *Vilokanam.* By observation.
13.	गुणितसमुच्चयः समुच्चयगुणितः। *GunitaSamuccaya Samuccayagunitah.* Product samuchya Samuchya Product.